A
GEOGRAPHY
OF
CHINA

A GEOGRAPHY OF CHINA

T. R. TREGEAR

Routledge
Taylor & Francis Group

LONDON AND NEW YORK

First published 1965 by Transaction Publishers

Published 2017 by Routledge
2 Park Square, Milton Park, Abingdon, Oxon OX14 4RN
711 Third Avenue, New York, NY 10017, USA

Routledge is an imprint of the Taylor & Francis Group, an informa business

Library of Congress Catalog Number: 2007026454

Library of Congress Cataloging-in-Publication Data

Tregear, T. R. (Thomas R.)
  A geography of China / T. R. Tregear.
    p. cm.
  Includes bibliographical references and index.
  ISBN 978-0-202-30999-6 (alk. paper)
    1. China—Geography. I. Title.

DS711.T68 2007
915.1—dc22

2007026454

ISBN 13: 978-0-202-30999-6 (pbk)

Preface

This book is intended primarily for serious students of geography but it is hoped that it will appeal to the general reader. For this reason technical terms have been used as sparingly as is consistent with correct meaning.

It has been the aim of the author to present, wherever the subject matter permits, a picture of geographical growth and to show the interaction of geographical environment and the human activity and institutions. An understanding by the rest of the world of what is happening in China and why it is happening is a matter of great urgency. Although present changes are taking place with breath-taking rapidity, they must be seen against the geographical and historical background if they are to be properly understood. A knowledge of the environmental facts is essential to an appreciation of the political, economic and social problems facing the Chinese people today. It is hoped that what is presented in the following pages will help towards this understanding.

The book has been arranged in four sections: Physical, Historical, Economic and Regional. Such a treatment necessarily involves some repetition but this, in some ways, is an advantage as it serves to emphasize the fact that the sections are themselves very closely interlocked. No apology is needed for including a considerable historical geographical section. The subjects treated under this heading are a few choice plums selected from a basketful of excellent fruit. No attempt has been made at a connected historical survey or treatment.

One aspect of Chinese geography which is fraught with difficulty is place-names. Cities, large and small, throughout history have changed their names many times. This habit continues today. For example, Anlu, a city on the Han River where the author lived in 1923, is now known as Chungsiang, whilst its nearest neighbouring city, which was then called Teinan, has now assumed the name of Anlu. This is all very confusing. In order to meet this difficulty, place-names used in the text have been plotted, as far as possible, on accompanying sketch-maps and have followed those used in the new edition of *The Times Atlas*. The names in brackets are alternatives and are often in more general use. Another confusion, which bedevils place-names and many other things besides, arises from the many systems of romanization of the Chinese script which have been attempted, e.g. Lantow, Lantao, Lantau. Incidentally this problem of finding a satisfactory romanization of Chinese writing which would have the same universal acceptance and understanding throughout the land as their present 'characters' have, has yet to be solved.

Another difficulty is that of securing adequate and reliable statistics. Records of many aspects of Chinese life and economy have been kept over a longer period than in any other civilization. For example, there are records of floods and droughts reaching back to the early Han dynasty. These, like most of the statistics of the country right up to the present day, need to be read with great care and caution. What constituted a flood or a drought was largely a subjective measurement by a district or provincial governor, whose judgement was coloured too often by the effect of prosperity but more usually of poverty (and consequently lower taxes) he wished to produce. Statistics compiled by the Chinese Maritime Customs (1854–1931), for so long under the control of Sir Robert Hart, are among the most reliable. The present Chinese People's Government began to develop a reliable statistical bureau between 1949 and 1958. Its work then suffered a disastrous set-back at the hands of fanatics during the 'Great Leap Forward' when it gradually emerged that figures for that year were unreliable to a degree. It is now most difficult to assess the amount of reliance that can be placed on figures for subsequent years. Statistics quoted for the post 1949 period are taken directly or indirectly mainly from the journals *Peking Review* and *China Reconstructs* and, to a lesser degree, from *Far East Trade* and translations of mainland papers and magazines.

It will be noticed that Taiwan, Hong Kong and Macau have been included in the text as though they were legally integral parts of China. It would be ludicrous to exclude them from description on the ground that at the moment they are politically administered by governments other than that of the People's Republic. They are geographically parts of one whole and it would seem virtually certain that, sooner or later, they will come under the jurisdiction of a Chinese government. This is essentially a geographical work, but anyone who has had experience of communism will know that its emphasis is primarily political. Consequently political aspects will now and again protrude themselves.

A glossary has been compiled (see pp. 333–4), which will be helpful, since it will often give significant meaning to a place name, e.g. Peking (*Pei* – north; *king* – capital); Nanking (*Nan* – south; *king* – capital); Shantung *Shan* – mountain; *tung* – east); Shansi (*Shan* – mountain; *si* – west).

It is hoped that full acknowledgment has been made of all published material used. Reliable and tested information regarding recent development and growth is difficult to obtain. The author was fortunate in being able to re-visit China in 1956 and to observe many of the changes which were already well under way in 1951 when he left the country.

Contents

CONTENTS

PLATES

MAPS AND DIAGRAMS

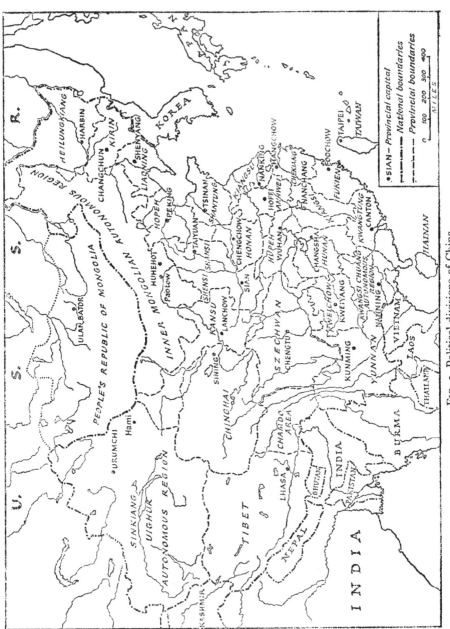

Fig. 1. Political divisions of China

Introduction

From the beginning an attempt must be made to convey to the reader some impression of the immense size of the country which we shall describe. It stretches from latitude 54° N. (Dzhalinda) to 18° N. (Hainan), and from longitude 74° E. (W. Sinkiang) to 135° E. (Khabarosk), but a mere recital of latitude and longitude, striking as it is, is rather sterile. It is perhaps more helpful to imagine the superimposition of China on other parts of the world better known to many readers. For example, if the China of the present People's Republic were superimposed on North America, its northernmost point, Dzhalinda in Manchuria, would lie somewhere in the vicinity of James Bay, while its southernmost point, Hainan, would coincide with Jamaica. The western border of Sinkiang would lie off the coast of California, while Khabarosk at the confluence of the Ussuri and the Amur would coincide with Cape Breton Island. For those who are more familiar with Europe a similar superimposition would show China covering the Mediterranean from end to end with Dzhalinda in the vicinity of Moscow and Hainan near Khartoum.

The area of the People's Republic of China is 3,657,765 sq. miles. This is some 620,587 sq. miles less than under the Manchu regime, owing to the secession of Outer Mongolia, when the estimated areas were:

	sq. miles
China Proper	1,532,800
Manchuria	363,700
Mongolia	1,367,953
Chinese Turkestan (Sinkiang)	550,579
Tibet	463,320
	4,278,352

This was a large secession. Economically it is of comparatively small importance since it is comprised largely of desert, semi-desert and poor steppeland, but politically and strategically it may have far-reaching implications.

For at least the last 2,000 years there has been a distinction in Chinese eyes between what is termed China Proper and the dependencies or colonies of Imperial China. China Proper embraces the eighteen provinces[1] which lie between the Pacific Coast and the Tibetan highlands in the west, and the series of ranges forming the edge of Inner Mongolia in the north. In these was developed the civilization which has had a continuous existence from ancient times until today. Outside have lain the 'barbarians' from whom, in times of Chinese ascendency, tribute was received. Due to its encirclement by great mountain wastes on the west, and mountain and desert on the north and north-west, China Proper has been singularly cut off by land from the influence of other civilizations. The dependencies of Tibet and Mongolia have acted as effective buffers. The only invasions experienced were from the nomadic peoples occupying the steppelands who were, in due course, either absorbed or expelled.

Similarly, the vast expanse of the Pacific Ocean for long militated against intercourse with the Americas to the east, regions which, in any case, were undeveloped. It is true that there was appreciable trading intercourse with the Arabs, Persians and Indians in early medieval times, but this had little real impact on so vast a land as China. There was no danger of invasion from this direction. Any such attack would have entailed far greater accumulation of capital and military 'know how' than then existed. Such penetration had to wait until the industrial revolution in the West had come to full fruition.

Thus it was that China remained in virtual isolation. Even in times of its greatest imperial expansion its contacts with Europe and even India were seldom, if ever, direct. It cultivated a rich civilization of its own, conservative and complaisant in its self-sufficiency. The stirring intellectual and spiritual revolution of the Renaissance, which so rocked Western thought, left China unmoved. The industrial and technical revolution of the eighteenth and nineteenth centuries made no appreciable impact on China until the twentieth century. China remained essentially rural and medieval in essence until very recently. The explosiveness of the sudden intrusion of Western ideas and techniques has therefore been all the greater. The rapid changes that have taken place have been attended by constant war, unrest and disorder, which were resolved only by the Communist revolution of 1949.

Although there has been this dramatic break-through in China's long period of isolation, the old parochialism has spilled over even into these revolutionary times. Anyone who has visited China in very recent years

[1] Hopei, Honan, Shantung, Shansi, Shensi, Kansu, Szechwan, Hupeh, Hunan, Anhwei, Kiangsu, Chekiang, Kiangsi, Fukien, Kweichow, Yunnan, Kwangtung, Kwangsi.

will have noted both the preoccupation with, and the self-satisfaction in, their own affairs, which characterizes so much of what they are now doing. Little interest is shown in what is happening abroad. Perhaps this is not unnatural in view of the colossal task which confronts the Chinese. Nevertheless it would seem to point to the fact that the influences of their long isolation are still present and that the Chinese still regard their country as

中 國¹, *Chung Kuo*, the Central Kingdom.

CHAPTER ONE

Physical Geography

'The physical features of a country represent the sum total of the constructive, deformative and erosive processes that have operated in that country throughout geological time.'

J. S. Lee

STRUCTURE AND RELIEF

The physical physiognomy of a country, its mineral resources, its vegetational cover and even, to some extent, the character of its people, are determined, in greater or less measure, by its geology and its structural history. It behoves us, therefore, to turn our attention at this early stage to the evolution which the topography of present-day China has undergone, to pass in rapid review the main geological successions and to observe the effect of the chief periods of revolution.

Much of the vast country is still unexplored or quite inadequately surveyed. A great deal of work is now being done in the field in China but, until this has been properly plotted, analysed and published, we must continue to rely on the work of such pioneers as Richthofen, Willis and Grabau, and of their successors, such as Andersson, Pierre Teilhard de Chardin, Young and Lee. It is to be hoped that in a few years time many of the problems and uncertainties that now exist will be cleared up.

The Ancient Floor

Underlying the whole of China, from east to west and north to south, is a floor of ancient rock. Cores of pre-Cambrian rocks lie exposed in many parts, mainly in ranges to the north of the Yangtze, in Inner Mongolia, Shensi, Shansi, Shantung and Liaoning. They are of two systems: the *Wutai*, named after the Wutai Shan in the Peking Grid, which is composed of highly metamorphosed sedimentaries (gneiss and schist), together with acid and igneous rocks; and the *Sinian*, first named by Richthofen and included in the Palaeozoic by Grabau, which is also sedimentary but much less metamorphosed.

I

Rising from this ancient floor are three massifs, none of which has been completely submerged by subsequent transgressions. The largest and most stable of these is Tibetia, whose pre-Cambrian ranges rise through thin layers of gravel, loess and sand. Gobia, although covered in parts during Tertiary times by seas which laid down clays and sands, has also been generally above sea-level. The third, named by Grabau as Cathaysia, is the south-eastern area of China and is the western remnant of a massif, which was raised by folding in the Sinian and existed throughout Palaeozoic times. The eastern part of Cathaysia, through violent folding and down-faulting in recent geological time, has been submerged, leaving Japan, the Philippines and festoons of islands marking its eastern rim.

Palaeozoic

A geosyncline, known as the Cathaysian geosyncline, running from the north-east in present Manchuria to the present mouth of the Ganges and thence westward in the Sea of Tethys, existed almost continuously through-out the Palaeozoic. During the Cambrian this geosyncline underwent considerable subsidence, while the surrounding land was peneplained. Even Shantung was partially submerged.

In Middle and Late Ordovician times, this geosyncline was interrupted

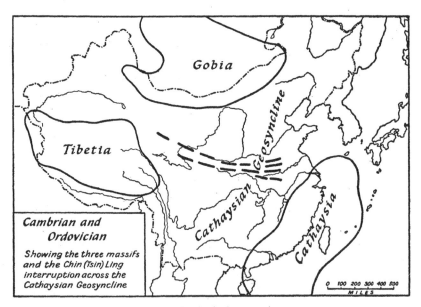

FIG. 2a. Geological succession

2

by the raising of a land barrier running east to west along the line of the present Chinling (Tsinling) in a movement known as the Hsiayuan. This east-west axis has been maintained to the present day and is one of the great geographical divides of China. In these same Ordovician times there was a general subsidence to the south, accompanied by intense folding in Indo-China. The area to the north of the east-west Chinling (Tsinling) axis, which had previously been subjected to repeated submergence, was now uplifted and so remained until late Carboniferous and Permian times, subject to a long period of erosion. In fact, this northern area has never again been deeply submerged.

Owing to this uplift and consequent absence of sedimentary rocks of the period, there is a lack of evidence of the Caledonian movement north of the Chinling (Tsinling). In the far north, the fold mountains of Sayan and Irkutsk on the east and west of Lake Baikal were raised. There was also considerable folding in the south-west, and in the south-east the Nanling and south-east highlands were raised.

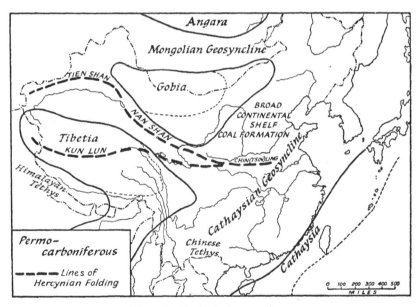

FIG. 2b. Geological succession

During Permo-Carboniferous times the sea (Chinese Tethys) spread northward, and great stretches of the north oscillated between land and sea, forming at times a broad, shallow continental shelf. It was in this period that the main coalfields of the north and centre were formed.

3

Towards the end of the Permian, the Tungwu orogeny occurred. It was a period of intense folding, corresponding to the Hercynian in the West. The great systems, comprising the Tien Shan, Kunlun, Altyn Tagh, Nan Shan and Chinling (Tsinling) were formed at this time, and their line continued through Shantung and the East Manchurian Mountains. The coal measures, previously laid down in the Carboniferous and Lower Permian, were greatly disturbed. During this revolution Tibetia remained firm and most of north China remained above sea-level, but considerable areas of Gobia were temporarily submerged. Most of southern China and Kweichow were also below sea-level.

Mesozoic

During the Early and Middle Triassic, south-west China, the Red Basin of Szechwan and the lower Yangtze valley lay below the sea. Thin bedded limestones, which are oil-bearing, were laid down at this time. During later Triassic times the greater part of Burma, Thailand and Indonesia were submerged as the Chinese Tethys rejoined the Himalayan Tethys from which it had been cut off during the Permian.

Subsequently, in the Jurassic, coal-bearing shales and sandstones, in

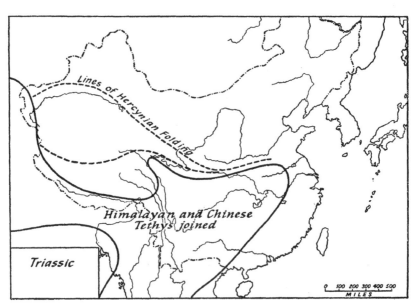

Fig. 2c. Geological succession

4

widely distributed basins of greatly varying size, were deposited.[1] One of these was the present Red Basin of Szechwan. North China for the most part remained above sea-level and was eroded to a peneplain, while the geosyncline to the south of the Chinling (Tsinling) continued to develop and deepen. In late Jurassic or early Cretaceous, the Yenshan or Ning-chinian movement occurred. This again was a period of extensive and

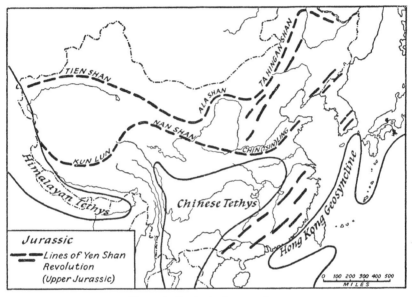

FIG. 2d. Geological succession

powerful folding. Its main lines were along the former Hercynian lines of compression, i.e. Tien Shan, Kunlun, Nan Shan and Chinling (Tsinling) and were continued north and north-east through the Ala Shan and In Shan on the one hand and the T'ai Hang Shan and Luliang Shan on the other to the Great Ch'ingan and Little Ch'ingan. Farther east and south it effected the shattering of the Shantung-Liaoning platform and heavily folded the land of south-east China (Cathaysia), Indo-China and North Borneo. A new geosyncline, known as the Hong Kong and Nippon Bays, was formed to the south-east extending from south Japan to Hainan.

Following this powerful revolution came a period of quiet sedimentation in continental basins, probably undergoing continual subsidence. The most notable and extensive of these basins were in Mongolia, where dinosaur

[1] J. S. Lee, *Geology of China* (London 1939), pp. 155–61.

5

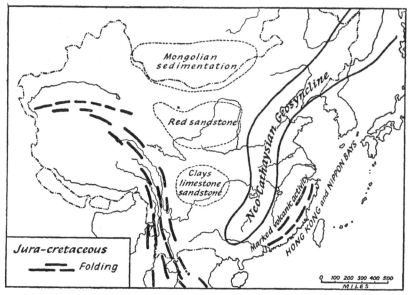

FIG. 2e. Geological succession

eggs of Middle and Upper Cretaceous were found by P. Chapman Andrews, and in the Red Basin, where clays, sandstones and limestones to a thickness of over 3,000 metres were laid down on the Jurassic coal measures. In north-west Shansi, north Shensi and Kansu beds of red sandstone were deposited. Lee comments on the oil-bearing potentialities of these Szechwan and northern deposits, potentialities which have been amply fulfilled by recent discoveries.[1]

This period of quiescence was followed by one of great volcanic activity everywhere outside the Cretaceous basins. The igneous intrusions of granite, granodiorite and porphyry at this time are largely responsible for the rich deposits of wolfram, tin, antimony, lead, zinc and copper, which occur especially in the southern half of the country.

The final phase of this Yenshan movement consisted of further folding and volcanic intrusion in Upper Cretaceous times, known as the Chinganian movement corresponding to the Laramide of North America, when the Rocky Mountains were formed.

Cainozoic

In mid-Tertiary times there was some marked folding in Sikang (Chamdo

[1] J. S. Lee, op. cit., pp. 183–4.

6

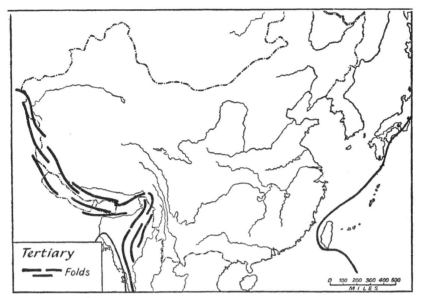

FIG. 2f. Geological succession

Area) but China, as a whole, bears few marks of the Alpine revolution which is so outstanding in Europe, western Asia and northern India. Apparently it affected only the extreme western margins of Yunnan and Szechwan.

In Pleistocene times at a period equivalent to the Würm glaciations the main deposits of loess from the north-west were laid down in Shensi and Shansi. The extent and nature of this fine, wind-borne soil are dealt with fully in a later chapter. Suffice it here to say that deposition is still going on, although apparently at a slower rate than formerly.

Quaternary

It is generally agreed that there has been a considerable uplift in the west during quaternary times, a movement which is continuing in the present day. In support of this contention Lee says: 'A striking case . . . In the neighbourhood of Tatsinlu and on the eastern side of the Minyu Gongkai (Minya Konka), Heim has actually observed the fluvio-glacial water cutting its own deposit formed in the glacial age to a depth of 100 metres. There is clear evidence that the glaciers in the high mountains have retreated in recent geological time. The argument that such stream erosion might be due to increased precipitation in post-glacial time can, therefore,

7

be safely ruled out.'[1] Further evidence of this recent uplift is seen in the physiographical history of the Yangtze (see p. 230 ff.).

Glacial Action

Until recently the view that China was left untouched by the glacial epoch was firmly held. Andersson, in referring to the loess deposits, says that in north China and Mongolia everything indicates that the climate in those parts was too dry to permit the existence of an ice cover.[2] However, in the Tatung Basin and the T'ai Hang Shan of Shansi, there are indications of glacial action such as erratics on the tops of high hills, striated boulders and U-shaped valleys. Lee suggests that local glaciers may have been formed towards the end of the Sanmen period when there was a fall in temperature and a rise in humidity.

Similar evidence of grooved rocks, parallel striation of rocks in both mountain and plain, cirque formations, boulder clay and U-shaped valleys in the Yangtze valley, notably in the Lushan Hills, Kiangsi, have, for many years, suggested that there was considerable glacial action in this area.[3] Doubt was cast on the validity of evidence in the form of parallel scratches since it was contended that, in view of the heavy weathering to which the rocks have been subjected, the scratches could not have survived. However, in 1936, the occurrence of unusually low water around Lake Poyang provided a unique opportunity for observations of formations, and it is now considered as conclusively established that there was widespread glacial action in this area.[4]

Topography

With this geological background in mind, it is now possible to take a general view of the relief and drainage of China. No attempt will be made to describe this in any detail, which would be a long and tedious undertaking. A more intimate treatment will be made in the regional section.

I. PLATEAUS AND BASINS

Western and north-western China consists largely of great upland regions, separated from each other by massive mountain systems. While

---

[1] J. S. Lee, op. cit., p. 207.
[2] J. G. Andersson, *Children of the Yellow Earth* (London 1934), p. 130.
[3] J. S. Lee, op. cit., p. 395.   [4] ibid., pp. 398–9.

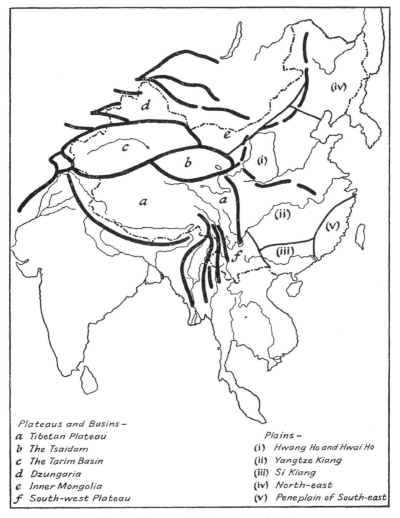

FIG. 3. Topographical divisions of China

these basins vary greatly in height from below sea-level to over 12,000 ft., they have one common characteristic. They are all inland drainage basins.

(*a*) *The Tibetan Plateau.* This great massif stands today at a general level of over 12,000 ft. Its boundaries are everywhere considerably higher than the interior: on the south the Himalayan ranges; on the west the Hindu Kush, which extend into the plateau itself in the Karakorams; on the north the Kunlun and Astin Tagh (Altyn Tagh); and on the east the north-south

9

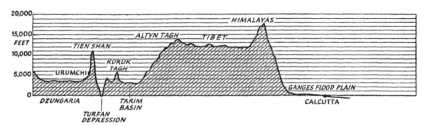

FIG. 4a. Section from Dzungaria to Bay of Bengal

folds of the mountains of Sikang. In its southern and eastern periphery great rivers find their sources and their way to the sea: the Indus and Brahmaputra; the Mekong, Yangtze and Hwang-ho. Most of the precipitation on the interior side of the boundary ranges flows inward and terminates in the many brackish lakes which stud the plateau. Innumerable east-west ranges, rising through thin layers of sand, gravel and clay, push their ancient cores to heights of over 18,000 ft.

(*b*) *The Tsaidam.* North-east of the Tibetan plateau is a smaller basin, the Tsaidam, which stands at a general height of about 9,000 ft. This is bounded on the north-west by the Astin Tagh (Altyn Tagh); on the north-east by the Nan Shan and on the south by the Kunlun. This, again, is an inland drainage area.[1]

(*c*) *The Tarim Basin.* North of the Tibetan plateau and at the much lower general level of 3,000 ft. lies the Tarim Basin. It is hemmed in by great mountain ranges; the T'ien Shan on the north, the Pamir Knot on the west and the Astin Tagh (Altyn Tagh) on the south. From these heights, glacier-fed streams descend, only to lose themselves in the sands and gravels of the Taklamakan desert, which occupies the centre of the basin and which is one of the most barren of the world's deserts. This area figures prominently in Chinese history, for it was along the line of oases which fringe the north and south edges of the desert, that the earliest contacts with the West were made.

To the north-east of the Tarim Basin and included within the same bounds is the smaller Turfan Basin, which is 940 ft. below sea-level.

(*d*) *Dzungaria.* North of the Tarim Basin is yet another inland drainage area, Dzungaria. The heart of this basin descends to heights of less than 1,000 ft. above sea-level. Thus we find three clear downward steps as we pass northward from Tibet to the border of the USSR. Dzungaria is enclosed

---

[1] P. Fleming, *News from Tartary* (London 1938).

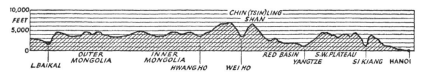

FIG. 4b. Section from Lake Baikal to Gulf of Tongking

by the T'ien Shan on the south; the Altai Mountains on the north-east, cutting it off from Outer Mongolia; and the Tarbagatai Mountains on the north-west, forming the frontier between China and the USSR.

Like the Tarim Basin, this area figures prominently in early East-West contacts, for one branch of the Silk Route passed along the northern edge of the T'ien Shan.

It should be noted that, while formerly Sinkiang and the Tarim were often used as synonymous terms, today Sinkiang is officially the political district embracing both the Tarim Basin and Dzungaria.

(e) *Inner Mongolia.* Inner Mongolia is the southern half of the Gobi, which geographically embraces both Inner and Outer Mongolia and is the great shallow, inland drainage basin enclosed by the Khangai, Altai and Nan Shan in the west; the Ala Shan and In Shan in the south; the Ta Ch'ingan in the east; and the Sayan and Irkutsk Mountains in the north. It descends to about 1,500 ft. in the centre. A number of streams, the greatest of which is the Estin Gol, descend from the Nan Shan, across the 'pan handle' of Kansu, to be lost in the heart of the desert.

(f) *South-west Plateau.* This plateau, which comprises the whole of Yunnan and the western part of Kweichow, is highly dissected. It is the only one of our classified plateaus whose drainage is entirely seaward.

2. THE GREAT PLAINS

The three east-west axes, along the In Shan, Tsinling and Nanling, noted by Lee,[1] have had marked influence on the three great rivers of China, determining their direction of flow.

(a) *The Hwang Ho* is confined between the northern axis of the In Shan and the central axis of the Tsinling. After it descends from the Tibetan massif, it meanders between these two until it finds an outlet of its own making across the North China Plain to the Gulf of Pohai. After cutting through the Nan Shan at the Liuchia Gorge, it turns north in a great loop and flows

[1] J. S. Lee, op. cit.

in a wide, shallow stream through the Ordos Desert. Turning south just north of Hokow, it cuts deeply through the loess, and passes through the region which cradled Chinese civilization, where it turns abruptly eastward at the confluence of the Wei River by way of the Tung Kwan gorge. Thence it emerges on to the wide alluvial flood-plain, which, although so subject to disaster, is yet so heavily populated.

(b) *The Yangtze Basin* lies between the Tsinling and the Nanling axes. Nearly one-third of the great river's upper course lies in the highlands of Sikang. After cutting its way through the Taliang Shan, it flows eastward through a series of former lake basins: first the Red Basin of Szechwan, with its four left bank tributaries; then the lake-studded Central Basin of Hunan, Hupeh and Kiangsi, drained by the Han, Yuan, Siang and Kan Rivers; and finally the 'delta' region of Anhwei and Kiangsu. These three basins form the most productive and most densely populated region in China.

(c) *The Si Kiang Basin.* Rising in the Yunnan plateau, the Si Kiang pursues an eastward course, south of the Nanling through Kwangsi and Kwangtung and enters the sea by an extensive delta at Canton. While not as extensive a plain as either the Hwang Ho or the Yangtze, it is important in that it has a sub-tropical climate and consequently is a double-cropping rice area.

(d) *The North-East (Manchurian) Plain.* This great undulating plain lies in the extreme north-east and is comprised of the three provinces of Heilung-kiang, Kirin and Liaoning. It is bounded on the west by the Ta Ch'ingan Mountains and on the north by the Hsiao Ch'ingan Mountains. The Eastern Manchurian Mountains form a considerable part of its eastern boundary, separating it from Korea. On the south the Jehol Mountains and the sea form the boundary. The plain is split into northern and southern halves by a low divide, the northern half being drained by the Sungari River and its tributaries, which empty into the Amur or Heilung Kiang, and the southern half by the Liao River.

3. THE SOUTH-EAST

Lying between the lower Yangtze and the lower Si Kiang basins is a region of much-folded mountains, having a north-east, south-west trend, the main ranges of which are the Wuyi Shan. Rivers are comparatively short and fast-flowing in steep-sided valleys down to the rugged south-east coast. It is a land which long resisted incorporation into the Chinese

Empire and, even today, it remains distinct, particularly in its dialects. Its people look largely to the sea for their livelihood.

CLIMATE

It is not surprising to find, in a country so vast as China, covering as it does so great a range in both latitude and longitude and including peaks of nearly 5 miles in height and basins below sea-level, that one encounters not one climate but a variety of climates. Here, in this vast sub-continent, are found hot desert climate in the Tarim Basin; cold desert in the Tibetan plateau; temperate continental climate in the north-east; the unique temperate Szechwanese climate of the Red Basin; and the tropical and sub-tropical in Kwangtung, to mention only a few. Yet, in spite of all these differences, close analysis will show that they are all variations on one basic theme of causation.

The theme is provided by the monsoon rhythm arising from the continentality of the huge land-mass of Asia. The variations on the theme, played on the instruments of varying altitude, contrasting ranges of seasonal and diurnal temperature, contrasting types of seasonal rainfall and precipitation and of quietude and storminess—all reveal in greater or less degree their causal relationship to this basic continentality, to a rhythmic seasonal change from a dense high pressure centre over the land in winter to a low pressure area in summer.

The Monsoon

The term *monsoon* is used in everyday conversation to connote a particular type of weather experienced in South-east Asia. Usually what is in mind is the onset of a summer season of heavy rain. It is true that this is a marked phenomenon of the monsoon, but it is the result of the monsoon rather than the monsoon itself, which technically is a wind system with a clear seasonal change in direction. In India this change is from south-west in summer to north-east in winter. In China the change is from south-east in summer to north and north-east in winter. To a less marked degree there is a similar seasonal change in eastern North America with an inflowing south-east wind in summer and an outflowing north and north-west wind in winter. However, owing to the fact that the land mass of North America is not nearly so great as that of Asia, the seasonal reversal of the winds is less marked. Moreover the general trend of highland is N–S in North

America, admitting an easier flow of warm and cold air than in East Asia, where the mountain trend is rather E–W.[1]

The Winter Monsoon

Asia is the greatest of the world's land masses and, in consequence of this, its heart experiences greater variations in temperature between summer and winter than any other part of the world. As autumn progresses, the land mass quickly cools in marked contrast to the slower cooling ocean along its southern and eastern borders. A great pool or mass of cold air settles in the large, shallow basins of Siberia and Mongolia, forming an anticyclone or centre of high pressure, which rises to 30·6 in. By November this cold air-mass is fully established: its centre covers the whole of Inner and Outer Mongolia, and spreads steadily southward until it meets the warm air-mass of the North Pacific Trades along a front rather to the south of the China coast and Taiwan in what is now known as the *West Pacific Polar Front*. This air-mass is a fairly shallow layer of less than 10,000 ft. It remains stationary over East Siberia and Mongolia from November to March and is generally in undisputed control.

From this great anticyclone there is an outflow of dry northerly and north-easterly winds over China bringing prolonged and bitter winters to the northern part of the country and really cold weather as far south as the Central and Lower Yangtze Basins. The barometric gradient between high pressure and low pressure is steep, and the resulting winds are strong.

Occasionally this bitter weather is interrupted by a welcome warm spell brought about, apparently, by depressions moving in from the west in the upper layers of the atmosphere. At 10,000 ft. the circumpolar westerlies move freely above the lower colder layer as shown in Fig. 5. F. K. Hare says of these disturbances: 'Though they invariably weaken as they cross the continent, not a few of these European or Atlantic storms actually penetrate to Manchuria or North China every winter. At such times the Siberian high, usually regarded as permanent, is displaced far to the north-east, or may even be absent . . . During the lull it is common for warm moist air (either mT or modified cP) to spread far northward across China and Japan, giving a welcome relief from the cold of the monsoon.'[2]

These lulls are of short duration and, as the cold air moves southward once more with the re-establishment of the high, further cyclones are

[1] See F. K. Hare, *The Restless Atmosphere* (London 1953), Chapter 12.
[2] ibid., p. 144.

14

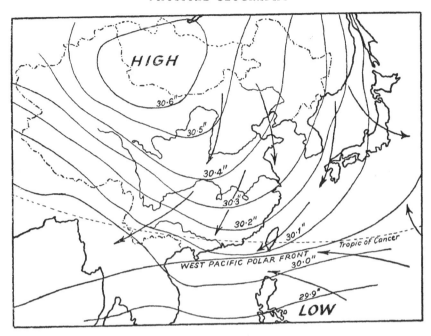

Nov - Mar

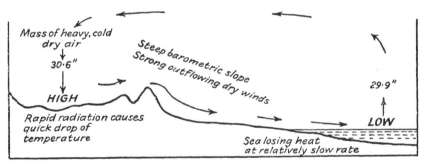

FIG. 5. Pressure and winds: winter

developed along the cold front and travel from west to east across China, continuing over Japan, often with increased intensity. These disturbances may arise on latitudes as far north as Dzungaria and Inner Mongolia, but their main development is along the Yangtze Valley, where the increased humidity may cause heavy snowfalls, hail and glazed frosts.

There is, however, a marked difference between the weather of north and south China at this time. Whereas the north experiences generally

15

clear, bright days with deep frost and occasional light snowfalls, there is more cloudiness and precipitation south of Tsinling Shan.

TRANSITION FROM WINTER TO SUMMER MONSOON

As spring (April) approaches the passage of cyclones increases both in number and intensity. The cold, heavy air-mass over the heart of the continent begins to warm up, at first very gradually, giving place to a continental low, while a high pressure centre is slowly established over the eastern Pacific. Wind direction is slowly reversed, northerly and north-westerly winds give way to those from the south and south-east, and a great current of warm, humid air moves in from the south. Its onset is felt earliest in the south and south-east. By May much of South China experiences mainly south and south-east winds and an increase in rainfall along the West Pacific Polar Front, which steadily retreats northward.[1] T'u Chang-wang comments on the gradualness of the onset: 'In India the "burst of the monsoon" can be followed pretty easily, but in China we have only a "burst of the cold waves", that is the winter monsoon, whereas the advance of the summer monsoon is very gradual and inconspicuous, hence the task of following its advance or retreat is more complicated.'[2]

By June the West Pacific Polar Front is over the Yangtze Valley, and by July it lies somewhere across the loess region and the Ordos, bringing occasional heavy rainfall to this dry region. Winds are now south-easterly over the North China Plain and southerly over Central and South China. The summer monsoon has now arrived. As can be seen from the map (Fig. 6) the barometric gradient between high and low is much gentler than in winter and consequently winds are lighter. The characteristic weather of the summer monsoon over much of China is hot, calm days of high relative humidity, which are very oppressive. Exasperatingly, any slight breeze which may have been blowing during the day so often fails in the evening, giving long, breathless nights. Coastal regions, especially in the south and south-east, are periodic victims of disastrous typhoon winds. Further inland very strong winds, known as *feng pao*, occasionally disturb the prevailing calm.

The duration of the summer monsoon varies between north and south, being shorter in the north on account of the lateness of onset. T'u Chang-wang and Hwang Sze-sung make a general estimate:

[1] Tao Shih-yen, 'Surface Air Mean Circulation over China', *Academia Sinica*, Vol. 15, No. 4, 1948.

[2] T'u Chang-wang and Hwang Sze-sung, 'The advance and retreat of the summer monsoon in China', *Meteorological Magazine*, Vol. 18, 1944.

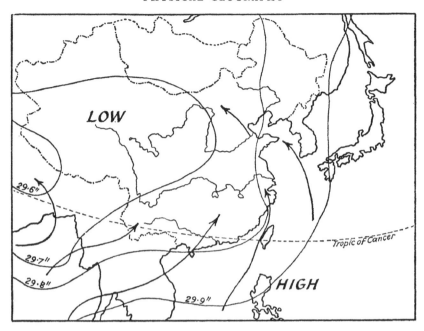

Apl – Oct

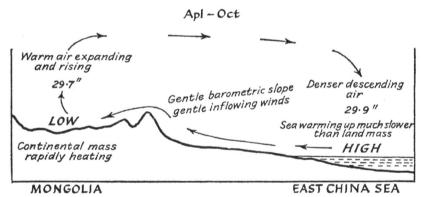

FIG. 6. Pressure and winds: summer

			Rainfall
S. of Nanling Hills	Mid April–end October	6½ months	75–90%
S. of Yangtze	May–mid October	5½	60–80%
N. China and N.E.	June–end August	3	60–75%
Kansu and Inner Mongolia	June–end August	3	less than 60%

Temperature

All three of the major influences on temperature namely altitude, latitude and land-mass play very outstanding parts in China. There is a difference of over 30,000 ft. between Everest and the Turfan basin. There is a difference of nearly 40° in latitude between Hainan (18° N.) and Moho (53° 33' N.) on the Amur in Heilungkiang. China has an area of 3,657,000 sq. miles and lies on the eastern side of the greatest land-mass in the world into which it penetrates some 3,000 miles. Small wonder therefore that we find vast differences both in range and level of temperatures from place to place and from season to season.

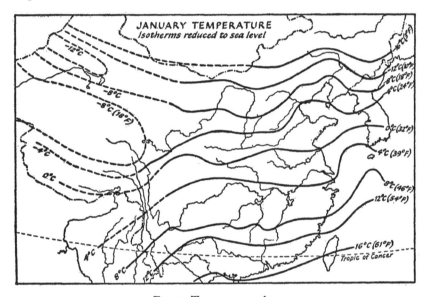

FIG. 7. Temperature chart

In winter there is a large and steady fall in temperature from south to north. Hong Kong has a January temperature of 60°F., Harbin —2°F. It will be noticed that the isotherms are remarkably regular in their alignment E–W. The isotherms bend slightly northward over the China Sea, but the ocean has very little moderating effect on winter temperatures since the winds are outflowing from the central low pressure and the cold Kamchatka current hugs the coast in its southward flow. Chefoo, on the Shantung coast, is as cold as Peking, both having an average January temperature of 23°F. Shanghai, on the coast, is as cold as Hankow, which is 600 miles inland.

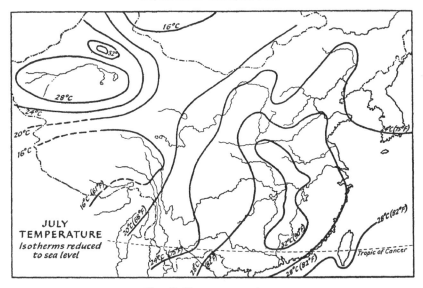

FIG. 8. Temperature chart

In summer the isotherms run N–S and reflect the influence of the in-flowing winds of the summer monsoon. There is a remarkable uniformity of temperature in summer, there being a difference of only 15°F. between places in China proper as compared with some 60°F. in winter. Everywhere it is hot and humid until one passes north-westward into the desert and semi-desert regions of Sinkiang, Mongolia and Dzungaria. Here annual temperature ranges are very great indeed. Luchun, which is in the Turfan basin and is 56 ft. below sea-level, has an annual range of 76°F. Compare this with Hong Kong, on the coast and in the tropics, which has an annual range of only 22°F.

As a result of continentality the cold pole of the northern hemisphere lies in the heart of Siberia, over Irkutsk and results in the 32°F. isotherm extending farther south in China than in any other region in the northern hemisphere. In China it reaches approximately 33° N.; in Europe 43° N. and in North America 38° N. The following are average temperatures for January of places of the same latitude:

Chefoo	23·5°F.	—	Algiers	54·0°F.
Mukden	8·6°	—	Rome	44·6°
Shanghai	37·8°	—	Port Said	56·1°

MEAN MONTHLY AVERAGE TEMPERATURE (in °F.) AND
MEAN MONTHLY RAINFALL (in inches)

Town	Altitude (in ft.)	Jan.	Feb.	Mar.	Apr.	May	June	July	Aug.	Sept.	Oct.	Nov.	Dec.	Total
Hong Kong	108	60 / 1·3	58 / 1·8	63 / 2·7	71 / 5·3	77 / 12·0	81 / 15·8	82 / 14·4	81 / 14·0	80 / 9·7	76 / 5·1	69 / 1·7	63 / 1·1	85·2
Canton	49	56 / 0·9	57 / 1·9	63 / 4·2	71 / 6·8	80 / 10·6	81 / 10·6	83 / 8·1	83 / 8·5	80 / 6·5	75 / 3·4	67 / 1·2	60 / 0·9	63·6
Swatow	13	59 / 1·4	57 / 2·5	62 / 3·1	70 / 5·6	77 / 9·0	82 / 10·5	84 / 7·8	83 / 8·4	82 / 5·5	76 / 2·9	68 / 1·6	62 / 1·5	59·7
Foochow	66	53 / 1·8	52 / 3·8	56 / 4·5	64 / 4·8	72 / 5·9	80 / 8·2	84 / 6·3	84 / 7·2	76 / 8·4	67 / 2·0	59 / 1·6	50 / 1·9	56·5
Kweiyang	4,560	37 / 1·0	42 / 1·1	53 / 0·9	63 / 2·9	71 / 7·0	72 / 8·1	76 / 9·0	77 / 4·1	68 / 5·4	57 / 4·2	53 / 2·0	47 / 0·6	46·3
Kunming	5,940	48 / 0·5	51 / 0·5	60 / 0·6	66 / 0·7	70 / 3·8	72 / 6·1	70 / 9·4	70 / 8·2	66 / 5·4	63 / 3·6	56 / 1·7	49 / 0·6	41·1
Shanghai	33	38 / 2·0	39 / 2·3	46 / 3·4	56 / 3·7	65 / 3·6	73 / 7·4	80 / 5·9	80 / 5·7	73 / 4·7	63 / 3·1	52 / 2·0	42 / 1·3	45·2
Kiukiang	110	40 / 2·5	42 / 3·3	50 / 5·9	62 / 7·1	71 / 6·8	79 / 9·6	86 / 5·6	84 / 5·2	77 / 3·5	66 / 3·8	55 / 2·7	45 / 1·7	57·7
Hankow	118	40 / 1·8	43 / 1·9	50 / 3·8	62 / 6·0	71 / 6·5	80 / 9·6	85 / 7·1	85 / 3·8	77 / 2·8	67 / 3·2	55 / 1·9	45 / 1·1	49·6
Changsha	295	43 / 1·8	46 / 3·8	51 / 5·8	63 / 6·1	71 / 7·8	79 / 8·8	86 / 4·8	86 / 5·2	77 / 3·4	66 / 3·6	55 / 3·1	43 / 1·8	56·0
Ichang	164	42 / 0·8	45 / 1·2	53 / 2·1	64 / 4·0	72 / 4·8	79 / 6·1	84 / 8·3	84 / 6·7	76 / 4·0	67 / 3·3	56 / 1·4	46 / 0·6	43·1
Chungking	755	47 / 0·6	50 / 0·8	58 / 1·4	67 / 4·0	74 / 5·5	79 / 7·1	82 / 5·6	84 / 5·1	76 / 5·8	67 / 4·6	59 / 2·0	50 / 0·9	43·5

Station														
Chengtu	1,560	44 / 0·2	46 / 0·4	55 / 0·6	63 / 1·7	70 / 2·8	76 / 4·1	78 / 5·7	78 / 9·7	71 / 4·1	64 / 1·8	56 / 0·4	46 / 0·1	31·6
Tientsin	7	28 / 0·2	32 / 0·1	43 / 0·4	56 / 0·6	67 / 1·1	76 / 2·4	81 / 6·8	78 / 5·1	72 / 2·8	58 / 0·6	41 / 0·4	32 / 0·1	20·6
Peking	131	23 / 0·1	29 / 0·2	41 / 0·2	57 / 0·6	68 / 1·4	76 / 3·0	79 / 9·4	76 / 6·3	68 / 2·6	54 / 0·6	38 / 0·3	27 / 0·1	24·9
Chefoo	00	23 / 0·5	31 / 0·4	40 / 0·6	53 / 1·0	65 / 1·4	73 / 2·4	78 / 6·8	78 / 6·1	71 / 2·6	60 / 1·0	47 / 1·2	35 / 0·6	24·6
Taiyuan	2,592	17 / 0·3	26 / 0·0	39 / 0·4	54 / 0·3	65 / 0·6	73 / 1·7	77 / 4·9	73 / 3·4	63 / 1·6	51 / 0·6	36 / 0·0	22 / 0·1	13·8
Sian	1,095	33 / 0·4	39 / 0·3	50 / 0·6	63 / 1·7	75 / 2·0	82 / 2·8	86 / 3·3	82 / 5·0	72 / 1·6	63 / 1·6	44 / 0·3	35 / 0·4	20·0
Saratsi	3,078	5 / 0·1	15 / 0·2	32 / 0·3	47 / 0·3	64 / 0·9	69 / 1·8	73 / 3·9	70 / 2·9	57 / 2·0	44 / 0·6	24 / 0·1	5 / 0·1	13·5
Dairen	33	25 / 0·5	25 / 0·3	36 / 0·7	48 / 0·9	59 / 1·7	68 / 1·8	73 / 6·4	76 / 5·1	69 / 4·0	58 / 1·1	43 / 1·0	30 / 0·5	24·1
Mukden	144	9 / 0·2	14 / 0·2	28 / 0·8	46 / 1·1	60 / 2·2	70 / 3·4	76 / 6·3	74 / 6·1	62 / 3·3	48 / 1·6	29 / 1·1	14 / 0·2	26·4
Harbin	525	−2 / 0·2	5 / 0·2	24 / 0·4	42 / 0·9	56 / 1·7	66 / 4·1	72 / 5·8	69 / 4·2	58 / 2·2	40 / 1·2	21 / 0·4	3 / 0·2	21·5
Urumchi	2,969	5 / 0·2	8 / 0·1	19 / 0·2	48 / 0·4	62 / 0·5	68 / 0·6	73 / 0·2	70 / 0·3	60 / 0·4	40 / 0·6	24 / 0·2	12 / 0·2	3·9
Luchun	−56	14	27	46	66	76	85	90	86	74	55	33	18	
Kashgar	4,003	22 / 0·2	32 / 0·2	47 / 0·2	64 / 0·6	67 / 0·8	76 / 1·0	82 / 0·2	78 / 0·2	67 / 0·0	54 / 0·6	39 / 0·0	28 / 0·0	4·0
Uliassutai	5,364	−13	1	20	37	48	67	66	62	49	31	10	5	

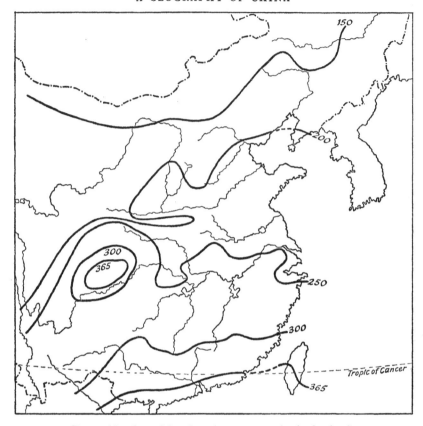

FIG. 9. Number of frostless days per year in the lowlands

Rainfall

Excluding the small south-west corner of Sikang, which shares the phenomenal precipitation of Assam, rainfall in China shows a general decrease from south-east to north-west, ranging from 85 in. in Hong Kong to 4 in. in Kashgar. A large part of south China receives over 60 in. per annum; the Yangtze basin between 60 in. and 30 in. There is a marked decrease in rainfall on crossing northward the Tsinling Shan, the most significant physical divide in China.

Everywhere the same seasonal distribution of a summer maximum and winter minimum is reflected. This is true even in remote Sinkiang, where rainfall is seldom over 5 in. a year. The lower Yangtze valley has a more even distribution of rainfall throughout the year than any other part, due

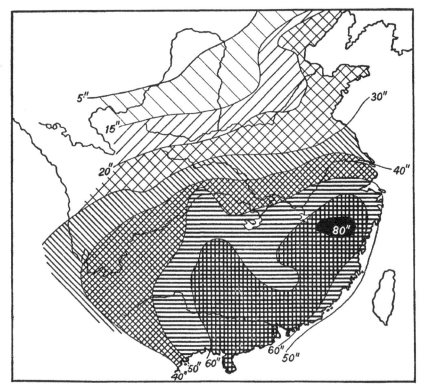

Fig. 10. Mean annual rainfall

to the passage of cyclones between the cold Polar and warm Tropical fronts but, none the less, a very clear summer maximum is maintained.

The basic cause of this seasonal variation is clearly the monsoon rhythm with warm, moist, inflowing winds in summer and cold, dry, outflowing winds in winter. However, this phenomenon by itself is not enough to account for the heavy summer rains. Rain does not always accompany the movement of humid air from sea to land. The increased temperature of the land in summer increases the air's capacity to hold moisture. In order that condensation shall take place, this warm, moist air must be cooled to saturation point. While in China this cooling is to some extent effected by the mountains, resulting in relief rain, the main cooling agent is the cyclone. Coching Chu comments strongly on this: 'It has always been taken for granted that the south-east monsoon in China, like the south-west monsoon in India is a rain-bearing wind. Yet the south-east wind in the eastern part of China is a dry wind in summer as well as in winter; and in

23

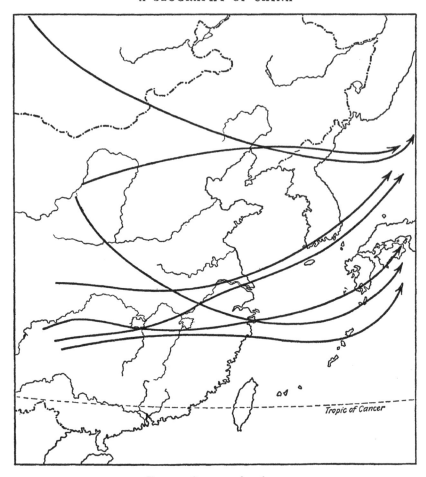

FIG. 11. Courses of cyclones

the Yangtze valley, when it blows consistently, drought is imminent. These facts were known to the ancient Chinese philosophers and one famous poet of the Sung dynasty wrote to the effect that, when the south-east wind blows, the rainy season is at an end. Recent observations confirm this statement. The apparent paradox is explained by the fact that the rainfall in China is mostly cyclonic in origin and not orographic as in India; and that most of the precipitation is in the cold sector. It is necessary to have a northerly or north-east cold air-current to lift the south-east monsoon to sufficient height for it to yield its quota of moisture.'[1] There is a tendency

[1] Coching Chu, *Journal of the Geographical Society of China*, Vol. 1, No. 1, 1934.

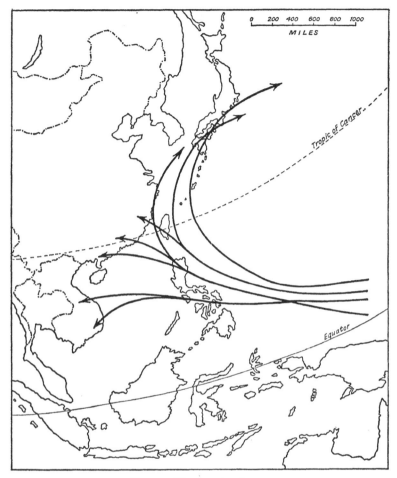

FIG. 12. Tracks of typhoons

for depressions to form over south-west China during the summer months. They move slowly along the Yangtze valley and it is these which bring the heavy rain to that area. It was the passage of a series of seven cyclones in July which caused the disastrous floods in the Yangtze valley in 1931. The nature and course of these temperate cyclones has been examined and plotted by Sung Shio-Wang.[1]

A further cause of rainfall is the typhoon or tropical cyclone which originates in the Pacific Ocean somewhere east of the Philippines. These

[1] Sung Shio-Wang, 'The Extratropical Cyclones of East China', *Memoirs of the National Research Institute of Meteorology*, No. 3, 1931.

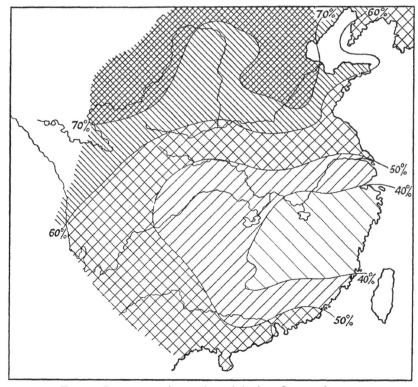

FIG. 13. Percentage of annual precipitation: June to August

typhoons move generally, but uncertainly, westward over the Philippines and north-west to the coast of south-east China. On an average there are eight typhoons a year, 80 per cent of which occur between July and November. They bring with them winds of over 100 m.p.h. and torrential rains, sometimes of over 20 in. in 24 hours. However, an examination of Hong Kong's rainfall records reveals the curious fact that years of serious typhoons do not coincide with the years of heaviest rainfall. These disturbances bring with them havoc and disaster to the coastlands where they strike. Happily they die out rapidly as they penetrate inland and thus they have little effect on the mainland as a whole.

Perhaps more important to mankind than the amount of rain that falls is the reliance that can be placed on the regularity of its fall. Variability of the annual precipitation from the mean increases in China from south-east to north-west, that is, in inverse ratio to the amount that falls. The area with least rainfall, i.e. the north-west, has a variability from the mean of 30 per cent; consequently this is the region notorious for famine. Even in

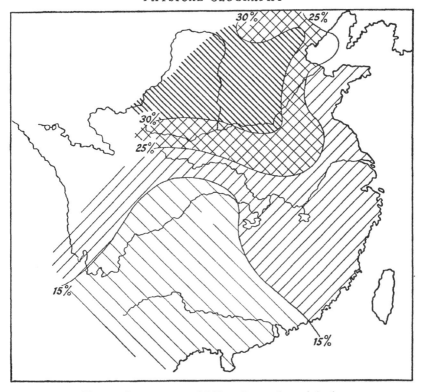

FIG. 14. Percentage variability of annual precipitation

the much more reliable south there is considerable variation, although, on account of the greater rainfall, its effects are not so serious.

Farther north in the Yangtze valley the yearly variation is more marked, as these figures of Hankow show:[1]

1916	1917	1918	1919	1920	1921	1922	1923	1924	1925
54·15	31·57	42·08	48·99	37·35	54·66	29·83	39·87	33·92	32·60 in.

Climatic Regions

Although the amount of published meteorological data is very limited for so large a country, there are enough stations of sufficiently long standing to warrant an attempt to divide China into broad climatic regions. During the last ten years a large number of weather stations have been opened. It is to be hoped that, in due course, their records will be published,

[1] S. V. Boxer, *Hankow Weather Guide* (Hankow 1926).

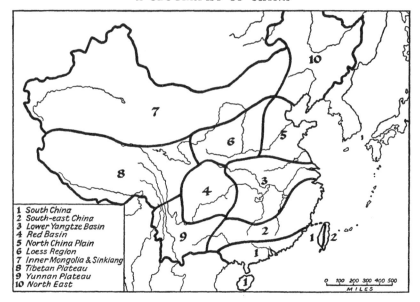

1 South China
2 South-east China
3 Lower Yangtze Basin
4 Red Basin
5 North China Plain
6 Loess Region
7 Inner Mongolia & Sinkiang
8 Tibetan Plateau
9 Yunnan Plateau
10 North East

FIG. 15. Climatic regions

when a more accurate and more detailed division will be possible than heretofore. The following is a general regional division:

I. HAINAN, SOUTH KWANGSI AND SOUTH KWANGTUNG AND THE WESTERN HALF OF TAIWAN

This region has an essentially tropical climate. The long summers are hot and the relative humidity is high, but there is a cooler, drier winter season, which brings some relief. Frost is known only occasionally on the highlands. The average temperature for January is over 55°F. The annual rainfall is between 60 and 80 in. with a very marked summer maximum. There is a twelve month growing period.

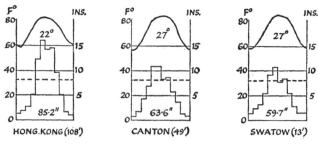

2. SOUTH CHINA; NORTH KWANGSI, NORTH KWANGTUNG, FUKIEN AND SOUTH CHEKIANG

This is a sub-tropical area. The summers are as hot as farther south and nearly as wet, but the winters are somewhat cooler. The coastal areas have a secondary maximum of rainfall in the early autumn, which is accounted for mainly by typhoons. The growing period is rather shorter, of about ten months. This is long enough for double cropping of rice but without a catch crop between, which is possible in 1.

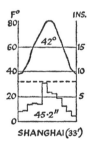
FOOCHOW (66')

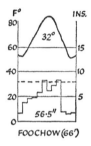
TAIPEI (30')

3. THE YANGTZE BASIN BELOW THE GORGES

This region has as hot a summer as 1. and 2. with a marked summer maximum of rainfall. Winters are much colder; heavy frost and some snow are brought with the bitter north winds which blow in January and February. The summers are oppressive, having both high temperature and high relative humidity. An unpleasant characteristic of the whole of the lower Yangtze basin is the consistent absence in summer of the evening breeze just when it would be most acceptable. The growing period is about nine months. Rice is grown in the south and central parts, but it faces competition by wheat and barley in the north.

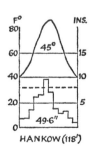
SHANGHAI (33')

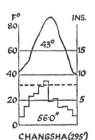
HANKOW (118')

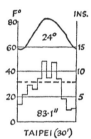
CHANGSHA (295')

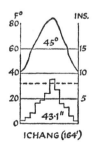
ICHANG (164')

29

4. THE RED BASIN OF SZECHWAN

This is a region with a unique climate. Situated in the heart of China as it is, one would expect it to have a climate of greater extremes than 3., but the reverse is the case. Chengtu, which is 1,000 miles inland from Shanghai, has an average minimum (January) temperature of 44°F. and an average maximum (July and August) temperature of 78°F. as compared with Shanghai, which has average temperatures of 38°F. and 80°F. for the same periods. The Red Basin is sheltered by the Tsinling and Ta Pa Shan from the bitter north winds and this accounts for the mild winter. The region is renowned for its humidity and mistiness. There is a Szechwan saying that 'when the sun shines, the dogs bark'. Rainfall is usually gentler than farther east and is ample for the temperature at various seasons. In consequence, there is a growing period of eleven months. It is a most productive region, having a greater variety of agricultural products than any other part of China.

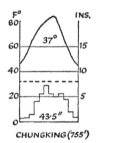

CHUNGKING (755')

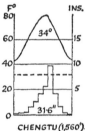

CHENGTU (1,560')

5. THE NORTH CHINA PLAIN

This also includes the peninsula of Shantung. Here the range of temperature between summer and winter is considerable, being at Peking

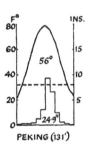

PEKING (131')

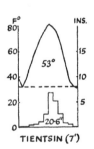

TIENTSIN (7')

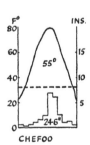

CHEFOO

56°F. as compared with Hong Kong's 22°F. Summers are as hot as, sometimes hotter than 1. and 2., although not nearly as wet. Summer rains are often torrensial and therefore of less value than they might be. Winters are long, cold and dry. The biting north and north-west winds descend on to the plain from the loess region and are often heavily dust-laden. The growing period here is not more than eight months.

6. THE LOESS REGION

As one moves north-west, so the continental nature of the climate becomes more marked. This is so in the Loess region. The range of average monthly temperature between summer and winter is, for most places, between 65° and 70°F. The winters are long and strong, bitter north-west winds sweep over the hills, whipping up heavy dust storms. There is a marked summer maximum of rainfall but it is sparse (10–15 in.), and what is worse, it is very variable and even more subject to cloudburst than 5. Hail storms can be very destructive. In years of drought the dust storms in summer are worse even than those of winter.

The sheltered valley of the Wei in this region is more favoured than the rest, having, as the graph of Sian shows, rather more rain and a more equable temperature.

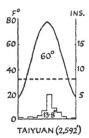
TAIYUAN (2,592')

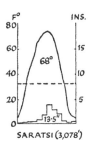
SARATSI (3,078')

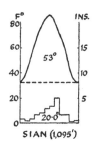
SIAN (1,095')

7. INNER MONGOLIA AND SINKIANG

These areas have temperate desert or semi-desert climates. The whole region, which is a series of basins in the heart of a land-mass, is subject to great contrasts in temperature between summer (80°–90°F.) and winter (5°–20°F.). There are also big diurnal ranges. Rainfall is very sparse and seldom more than 4–5 in. per annum on the lowlands, and moreover is very uncertain. This is an area of strong, often dust-laden winds.

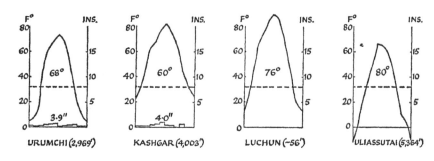

URUMCHI (2,969')   KASHGAR (4,003')   LUCHUN (-56')   ULIASSUTAI (5,364')

8. THE GREAT HIGHLAND PLATEAU OF TIBET

This forms a separate division. A great deal of it lies at over 14,000 ft. In consequence the density of the air is much less than on the lowlands. The air is dry and clear, with resulting rapid insolation and radiation. Both seasonal and diurnal ranges of temperature are great. Rainfall for most of the region is negligible. The south-east corner of Tibet in which Lhasa is situated and the only part for which any reliable meteorological statistics exist, shares the monsoon climate of north-east India, as the accompanying graph shows, and is in no way typical of the climate of the plateau.

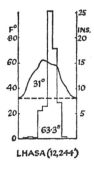

LHASA (12,244')

9. THE YUNNAN-KWEICHOW PLATEAU

This plateau is between 4,000 and 8,000 ft. in height. Its summer, in consequence, is not as hot as the regions 2. and 3. to the east. Its winter is not influenced by the out-flowing north winds from the continental high. There is a summer maximum of rainfall, which is ample for the latitude and temperature. The resulting climate is described, by those who have experienced it, as delightful and almost ideal. The deep north-south valleys of the headwaters of the Yangtze, Mekong and Salween, which lie to the west, however, are unhealthy and malaria-infested.

32

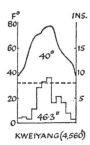

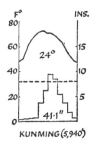

KWEIYANG (4,560) KUNMING (5,940)

10. THE NORTH-EAST OR MANCHURIA

This region has an east coast continental type of climate. Temperatures increase in seasonal range the farther north-east one goes. As a result, the growing period is shorter in Heilungkiang than it is in Liaoning. Rainfall also decreases to the north-east but is everywhere adequate for the growth of wheat and kaoliang.

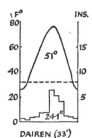

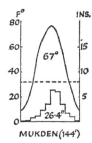

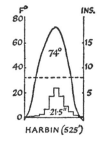

DAIREN (33′) MUKDEN (144′) HARBIN (525′)

SOILS

While the rocks of a region are the original basic source of all soil, it is the climate of that region, which, in the long run, determines the nature of its soil, aided by the organic agents, the flora and the fauna, which are themselves both cause and effect, and aided also by the action of human beings.

It is estimated that only about one tenth of Chinese soil is residual and that 90 per cent has been transported either by wind—witness the great deposits of loess in Shensi and Shansi—or by water in the great alluvial plains of north China, the Yangtze and Si Kiang. There is little or no evidence of glacial transportation on any large scale—certainly nothing comparable to that of the North German Plain or Canada. Human beings in China throughout the centuries have been no mean transporters: terracing, carrying soil from field to field and from river bed to field.

Soil formation depends less on the method of its origin and the nature of the parent rock than on the type of climate to which it is exposed. Both temperature and rainfall play leading parts and therefore it is not surprising to find that there are marked differences between north and south China. The north, being both colder and drier, tends to form its soils mainly by mechanical action, i.e. by frost and exfoliation. Because the climate is cold, chemical action is weak and soil formation slow. Because it is dry, there is little leaching of the soil. Thus the soils of the north are generally slow to come to maturity but tend to retain their fertility for a much longer period than the lands of the south, where macro-climatic conditions are reversed. In tropical and sub-tropical regions, where temperatures are much higher and rainfall much heavier, chemical decomposition of the rocks is much quicker. Through the combination of these two, the higher temperature and heavier rainfall, granites of Kwangtung and Hong Kong break down remarkably quickly and to astonishing depths. The heavy rainfall causes rapid and severe leaching. Thus the soils of the south tend to come to maturity quickly but are liable to lose their fertility very rapidly. This difference between north and south, and particularly the ease with which the soils of the south lose their fertility, is of the highest importance in considering the planning of the economic development of the country. We therefore must look more fully into the subject.

Gourou, in his admirable little book, *The Tropical World* points out that 'tropical soils are poorer and more fragile than those of temperate regions. Great care is needed in using them if their further impoverishment and destruction are to be avoided.'[1] The luxuriant forests of the tropics and sub-tropics give the impression of great fertility. The use of migratory *ladang* cultivation of forest clearings in the tropics by primitive tribes and the fact that these clearings lose their fertility in about a couple of years is enough in itself to make us distrust such an impression. The fact is that these forests are just able to maintain a state of soil equilibrium. The leaf fall is just sufficient to maintain a supply of humus, and the leaf cover provides the very necessary protection of that humus against rapid destruction by the sun. Then, too, the forest cover helps to prevent rapid leaching and if this cover is removed the soil loses its fertility very rapidly. Any fertile soil met with in tropical or sub-tropical conditions will be found to be either one of

[1] P. Gourou, *The Tropical World* (London 1953).

recent origin, such as newly laid-down river or marine alluvium, or basic volcanic ash, or one which is receiving constant renewal by human action. The writer remembers vividly learning this lesson when he tried to demonstrate in Central China the superiority of the temperate western method of digging in manure over the local Chinese method of almost daily top feeding with very weak liquid night-soil. He found that his own plots maintained their fertility for only three or four months instead of the two years he had expected, while the top-fed plots continued to yield as long as the feeding was maintained. Manure, which in England would suffice for two or three years, may be leached away in a matter of months. The infertility of tropical and sub-tropical soils, unless constantly fed (as they are generally in south and central China) is reflected in the figures below:[1]

Average Yield of Rice per Acre 1926/27-1930/31

	Bushels per acre		Bushels per acre
Temperate Countries		Siam	50
Spain	187	Indonesia	47
Italy	122	Brazil	44
Japan	107	India	41
USA	65	Philippines	35
Korea	56	Malaya	35
Tropical Countries		Madagascar	35
Sierra Leone	62	French Indo-China	32

As a result of the difference of climatic conditions between north and south, of the drier, cooler north and the wetter, hotter south, there is a two-fold major classification of soils. Generally speaking the soils north of the line of the Nanshan, Chinling (Tsinling) Shan, Funiu Shan are pedocals—calcium soils from which the lime has not been leached—whilst south of this line are the pedalfers from which the lime has been leached, leaving much iron and aluminium and increasing in acidity as the tropics are approached.

Pedocals

These can be given a wide five-fold division:

1. *Desert soils*, which are generally grey or yellow-grey in colour, are found in the north-west, Kansu, Inner Mongolia and Sinkiang. They are

[1] P. Gourou, op. cit.

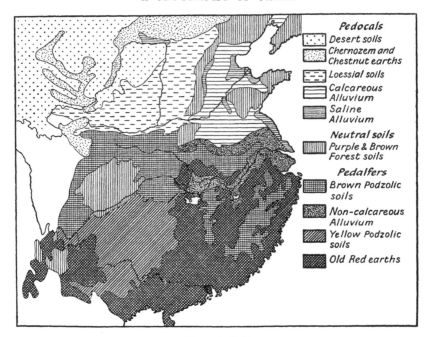

Pedocals
Desert soils
Chernozem and Chestnut earths
Loessial soils
Calcareous Alluvium
Saline Alluvium
Neutral soils
Purple & Brown Forest soils
Pedalfers
Brown Podzolic soils
Non-calcareous Alluvium
Yellow Podzolic soils
Old Red earths

FIG. 16. Soils

light in texture and liable to be blown away. It is probable that the loess farther east has been derived from this region. Generally the soil is thin, but given sufficient depth and adequate water supply by irrigation, it can be farmed. Considerable effort has been made to develop these soils in Sinkiang in the last decade. Great care has to be taken to meet the danger of alkali accumulation.

2. Lying roughly between the desert soils and the loess is a belt of *chernozem, chestnut* and *black earths.* These are light soils, developed in semi-arid areas and under natural grass. They are potentially very productive and have been a constant temptation to the Chinese colonist, luring him north-westward only to send him back once more after experiencing the short growing season and the variable rainfall. After a series of dry summers he can watch his fields, denuded of their natural grass cover by cultivation, whipped away eastward by the wind.

3. Covering an enormous area of Kansu, Shensi and Shansi is the famous *loess of North China,* by far the biggest deposit of its kind in the world. This fine, windborne soil has been transported from the desert soils and

deposited in beds of thickness, varying from 100 to 250 ft., masking much of the former relief.[1] The process is still going on. Being wind-borne, loess is generally unstratified and is laid down without the horizontal bedding planes of water-borne deposits. It is so fine a texture that it holds moisture easily, which is drawn to the surface by capillary attraction. A further characteristic is its vertical cleavage, giving steep, often sheer, valley sides. Given adequate water supply, loess is very fertile indeed. As we have seen, it is easy to work and it thus enabled Neolithic Man of China to move forward into the primitive agricultural stage.

4. Spread out over the whole of the North China Plain are the deep *alluvial deposits of the Hwang-ho flood plain*. An artesian well, bored to a depth of 2,840 ft. in 1936, revealed that the shells of the same small bivalved *Lutraria* and of small snails were present at the greatest depths as are present in the deposits being laid down today.

'The sand in which they were discovered had once been the top layer of the Yellow River delta . . . The digging of this well demonstrated, as no previous excavation had done, how much the rock floor of this part of the Asiatic continent has sunk and what an immense load of silt and sand the Hwang-ho has carried down and spread over the sinking plain in order to keep the ancient home of the Chinese above sea level.'[2]

The alluvial soils are generally light in texture and easily worked. They are mainly yellow or grey in colour. This is by far the most productive of the calcium soils, having both sufficient supplies of mineral foods and an adequate and reasonably reliable rainfall. Consequently there is a very dense population over the whole area. The soil has been worked and worked again over the centuries and has been most carefully tended, fed and replenished by the farmer. It is in regions such as this that it is difficult to determine whether nature or man is the chief soil-maker.

5. Scattered widely over north China are patches of *saline* and *alkaline soil*. The most extensive of these are (i) a long strip along the coast of Kiangsu, (ii) a considerable area around the mouth of the Hwang-ho, and (iii) patches in the flood plain of the river itself, and along the northern side of the great bend of the river. All these areas are associated with a combination of poor drainage and aridity, where the rainfall is insufficient both to dissolve and carry away the salts in solution. Sodium chloride, sodium sulphate and sodium carbonate are all present, but happily the latter is not

[1] G. B. Cressey, 'The Ordos Desert of Inner Mongolia', *Dennison University Bulletin*, Vol. 33, No. 8, Art. 4.

[2] N. Carrington Goodrich, *China* (University of California Press 1946), p. 42.

very prevalent. The present government is giving a good deal of thought to the reclamation of these saline areas.

Neutral Soils

6. *Purple and Brown Forest Soils.* Between the pedocals of the north and the pedalfers of the south lie, not unnaturally, soils which are neutral, i.e. where soil formation and leaching are approximately in balance. The purple and brown forest soils of the Red Basin of Szechwan come within this category. This is the land where 'the dogs bark when the sun shines', a land of mists and light rains, which do not cause rapid leaching. This balance between alkalinity and acidity is maintained in the higher lands of Honan, Hopei and Shantung, where purple forest soils also occur.

Pedalfers

Once south of the Tsinling-Fu Niu Shan line we are in the region of podzolic soils.

'Podzolization is understood to be a process by which soils are first leached of their easily soluble components and then the iron and manganese family are preferentially mobilized from the upper horizon. Laterization is thought of as the opposite process. The silica is preferentially mobilized and the iron and aluminium accumulate in the surface horizon as a residual concentration.'[1]

1. *Brown Podzolic Soils.* This reddish-brown soil is distributed widely over the highlands surrounding the west, north and east sides of the Red Basin and over the mountainous areas of Hupeh, Hunan, Chekiang and Fukien. It is less severely leached than the soils farther south and is therefore less acid. Formerly under forest, much of this has been stripped, with the result that the hill slopes have been eroded and the rice-wheat lands of the plains covered with a sterile clay. Over-enthusiasm to increase cultivation in recent years has led to unwise development of these soils, which are better kept under forest.

2. *Non-calcareous Alluvium.* The flood plain of the Yangtze below the Gorges, i.e. from Ichang to the sea, is overlain with a thick cover of non-calcareous alluvium, as also is the greater part of the Hwai basin. These soils are largely neutral, being neither calcareous nor acidic and for the most part are exceedingly fertile and of good texture.

[1] Carter and Pendleton, *Geographical Review*, October 1956.

3. *Yellow Podzolic Soils.* The plateau lands of Yunnan and Kweichow are characterized by strongly podzolized yellow soils of many types, all of which are strongly acid. The highland valleys are reasonably fertile but require constant feeding. The hillslopes, like all the southern regions, are best kept as forest lands.

4. *Old Red Earths.* These are the preponderating soils of southern China. They are increasingly acidic as the Tropic of Cancer is approached, when a distinct tendency to become lateritic can be detected. The only true laterite in China is found in Hainan. However, over large areas of Kwangtung and Kwangsi much of the hillsides are covered with a hard crust, an 'iron pan'. This may be only a fraction of an inch in thickness, but it is highly resistant. This iron pan, so essential in the paddy field when it is found some 12 to 18 in. below the surface, is fatal to cultivation when it appears on the surface. Gourou describes it as a 'pedological leprosy'. The Old Red Earths, like the Yellow Podzolic soils, are best left to their natural forest cover. The alluvial valley bottoms are cultivated, but because of heavy leaching they need constant feeding.

Soil Erosion

In view of his long attachment to the soil and his veneration for his ancestral lands, it is rather surprising to find how careless the Chinese peasant has been in preventing the loss of soil through erosion. Largely through ignorance he has 'left undone those things which he ought to have done and done those which he ought not to have done' and so imperilled the very thing on which his life depends. He has had a false sense of security. 'Land is there. You can see it every day. Robbers cannot take it away. Thieves cannot steal it. Men die but land remains.'[1] He has not been quick to recognize the many robbers, some of his own creation, that are ever ready to filch his land from him.

Soil erosion in all its forms can be found within Chinese borders, in some places to a disastrous extent. The sequence is the same no matter what the location. First the natural cover of the land is lost. The causes of this loss may be deforestation, unwise cultivation, overgrazing, cutting and burning off the grass and scrub, or desiccation from natural causes. Each and all these will lead to a loss of organic matter and with it the loss of structural stability. The valuable crumb structure breaks down, porosity is lost and the stage is set for rain to wash away the top soil by sheet and gully erosion or, in drier regions, for the wind to whip it away.

[1] Fei Hsiao-tung, *Peasant Life in China* (London 1947), p. 182.

Of all the regions of China, the loess region is the most vulnerable since it is subject to the action of both rain and wind. Its natural vegetational cover is grass and scrub, although it is maintained that formerly much was covered by forest. The most insidious form of attack is sheet erosion. During the wet summer when the ground is saturated, the soil is carried down even gentle hillslopes in innumerable tiny gullies or channels. The immediate effect is not easily noticeable and is masked by the next plough-ing. Nevertheless there is a tendency for a steady downward movement. An attempt has been made over many centuries to meet this danger by terracing, and the whole of the cultivated countryside presents a picture of large terraced fields on more or less gently sloping hills, each field being divided from the next by a low wall of sods. The method has been generally successful in preventing wholesale erosion, but neglect over even a short period, perhaps due to political unrest or to famine, will result in gullying from which recovery is well nigh impossible. Terracing here needs to be reinforced by scientific contour ploughing to make arable farming secure from erosion. In recent years the Puerto Rican method of digging pits a couple of feet deep all over the field and cultivating around them has been successfully practised in a few places. The pits serve to hold and conserve any heavy rainfall and prevent both sheet and gully erosion. It is not, however, a method which commends itself to large-scale farming.

Loess is very prone to gullying. The smallest depression, the slightest channel or foot-path may be sufficient to set in train disastrous erosion. We have noted that loess has a marked vertical cleavage and also that it has a high lime content. We have seen also that, although this is a region of sparse rainfall, the rain often falls in very heavy storms or cloudbursts. These three factors combine to produce very steep-sided, often vertical gullies or gorges, which may be deeper than their width and produce the characteristic dissected landscape of the loess.

Andersson describes yet another type of erosion in the loess:[1]

'During the great summer rains, it is true, the water rushes in cascades from the fields down into the depths, but the essential process is of quite a different kind. In order to understand the manner of formation of the loess ravines we must examine the yellow earth a little more closely. The typical Huang T'u is a greyish, yellow dust, which does not as a rule show any stratification, but shows on the other hand a remarkable capacity for adhering to perpendicular cliffs. This fine, porous earth easily lets through water which falls upon its surface. Consequently only part of the summer rains drains off its surface. For a large part of the rainfall it acts as a sponge, or, perhaps

[1] J. G. Andersson, *Children of the Yellow Earth* (London 1934), p. 129.

better, like a gigantic filter, through which the water sinks to the bedrock of the loess deposits, consisting of gravel, Tertiary clay and solid rock. The lower part of the loess soil in this way often becomes saturated with water and assumes a consistency like that of a thin porridge or gruel. This bottom layer then slowly begins to move and slides down any slope and in proportion as the saturated bottom slides away towards the open valley, the super-imposed, relatively dry mass of loess sinks down perpendicularly. This vertical movement may be studied everywhere in the ravines, in which one sees large and small blocks of the old vegetation-covered surface in all sorts of more or less inclined positions halfway or more down to the bottom of the ravine.'

It is the erosion of this loess, this *huang-t'u* that is the main reason for the constant flooding of the lower Hwang-ho, the flood plain of which is built largely of loess silt. Nothing short of a gigantic afforestation of large areas and a reversion to pasture of much of the rest will suffice to anchor this fine, light soil.

In south and central China the contrast between the meticulous care lavished on cultivated fields and the almost contemptuous disregard of the welfare of the hillsides and the rough land, has been most striking. So much of the countryside has been denuded of its natural forest cover to make room for cultivation in the valleys and lower hillslopes. Happily the careful terracing and cultivation of the rice fields has, to a large extent, afforded protection against erosion. However, nearly everywhere the higher hill-slopes have suffered from severe sheet and gully erosion. This is the result of generations of disregard of the elementary rules of soil conservation. The hills have been stripped not only of their trees but also of the grass which is cut annually and the very roots grubbed up for fuel, thereby constantly exposing the soil surface. The cooking stove of south China is designed to burn grass and wood. If an acceptable coal-burning stove could be designed and communications improved sufficiently to distribute the coal, which even in the centre and south is abundant, this would probably do more than any other measure to check erosion.

There is an almost universal custom of burning off the hillsides in the autumn each year. Apparently there are many reasons for this. One un-doubtedly is the historical fact that the Chinese farmer is an agriculturalist, that in earlier times the land was forested, and to clear this obstacle he felled and burned. The idea still holds. Other reasons put forward are that the clearance of the hillsides gives protection as it deprives bandits and wild animals of cover. There is still quite a lot of wild game—deer, wild pigs and even some leopards and tigers—in the uplands of China, which cause a lot of damage to crops.[1] During the unstable political conditions which have been prevalent throughout the whole of the first half of this century danger from robbers and bandits has been real. Two other reasons are given for the burning. One has a pseudo-scientific flavour, i.e. that potash from the burnt grass and brush is washed down and replenishes the rice fields below and also that the subsequent young grass on the hillsides is better grazing. Lastly, there is some truth in the statement that the peasants like to see the hillsides lit up at night. Whatever the reasons may be, there is no doubt that the practice of burning off the hillsides contributes to the dangers of erosion.

NATURAL VEGETATION

So much of China Proper has been so densely populated and subjected to the action of agricultural man so long, that it is difficult to know, in some parts, exactly what would grow if the land were left untouched. For example, most of the great North China Plain would probably be covered by deciduous woodland if uninhabited. As it is, it is now one vast patchwork of cultivated fields in which trees are a luxury.

Being itself largely the outcome of climate and soils, there is a close correspondence in distribution between these and the various types of natural vegetation. Before embarking on a brief survey of the distribution of vegetational types, mention should be made of the differentiation of growth on the north-facing and south-facing mountain slopes, which is noticeable throughout the whole country. The north-facing slopes tend to be more generously covered than the south. If the south-facing slopes have short grasses, those facing north are likely to have long and more abundant grass or scrub. If there is grass and scrub facing south, we may look for woodland on the north-facing slopes. This differentiation is due to the difference in insolation, particularly in the dry winter months when the south-facing slopes are subject to great evaporation.

In the north west, where there is not true desert, there are vast areas of very sparse xerophytic (drought-resisting) vegetation, within which, in the lower lying land and depressions, are patches of halophytes (salt-tolerant plants), notably in the Tarim Basin, the Ordos, the Tsaidam and the Gobi. Skirting the southern edge of the Gobi is a wide belt of grasses of many kinds and of varying abundance, the main home of Mongolian nomads.

[1] G. Fenzel, 'On the natural conditions affecting the introduction of forestry in the province of Kwangtung', *Lingnan Science Journal*, No. 7, June 1929.

The natural covering of the loess region is mainly grass and scrub, although there is some forest land on the higher hills, particularly where these penetrate above the loess deposits.

Spread out over the whole of the North China Plain, the Wei Valley, the lower Fen, the flood plain of the Yangtze below the Gorges and the flood plain of the Hwai, are the farmlands of China, which allow of little or no natural vegetation. A strip of salt-tolerant vegetation should be noted dividing the cultivated Hwai plain from the sea.

Stretching in a broad belt from the Nan Shan in the west, through the Tsinling and Fu Niu Shan, is a broad belt of coniferous and deciduous forest, the broad leafed trees increasing to the east and south. These forests continue in the Shansi highlands, which form the western border of the North China Plain and which in earlier days provided the hunting forests of the Shang nobility. The highlands of Shantung also have a similar forest cover.

From this belt southward to the sea the natural vegetation of China is forest of one kind or another. China enjoys a great variety of trees. There are twenty-six species of conifers in the country and a great variety of broad-leafed trees, including the *Eucommia*, the camphor, the nanmu

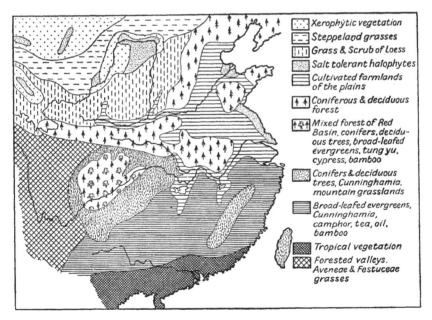

Fig. 17. Natural vegetation

43

(Phoebe), the sassafras, the t'ung yu (wood oil), tea oil, Chinese tallow and varnish tree. In the more populous south and east most of this forest has been stripped. A recent Russian survey estimates that, whereas China should have a forest cover of 30 per cent in order to give adequate control of flood and drought, it has, in fact, only about 5 per cent, which gives some measure of the deforestation which has taken place.

The southern forest lands can be divided generally into the following categories:

(*a*) Very mixed tree growth of the Szechwan Red Basin, which includes conifers and deciduous, broad-leafed evergreens, t'ung yu, cypress and bamboo.

(*b*) Surrounding the woodland of the Red Basin is a broad forest belt of more uniform character, consisting of conifers and deciduous trees, interspersed with mountain grasslands. A similar belt along the highlands of the south-east contains a great deal of *Cunninghamia*, as also do the highlands of Taiwan.

(*c*) The natural cover of the greater part of Hupeh, Hunan, Kiangsi, Kweichow Chekiang and Fukien is broad-leafed evergreen forest, a great deal of *Cunninghamia*, pine, camphor, tea, oil and bamboo.

(*d*) Kwangsi, the coastal regions of Kwangtung and the island of Hainan have tropical vegetation in considerable variety and, where untouched, in abundance. Evergreens of all kinds, pines, bamboos and large fleshy grasses, such as wild pineapple, all flourish.

(*e*) The valleys of the deep-cut ravines of West Szechwan and the Chamdo Area are densely forested with deciduous and coniferous trees, while the tops of the mountain ranges and the plateaus are covered with grasses of the *Aveneae* and *Festuceae* tribes.

CHAPTER TWO

Historical Geography

'Mature historical understanding requires full recognition of the factors of physical geography, climatic stimulus (where it can be proved) and the character of the environment as a whole; but it also demands an appreciation of the dynamics of social groups.'
                                                                    Owen Lattimore

It has been said that the trouble with Chinese history is that there is so much of it. Authentic written records carry us back well into the first millenium BC, and archaeological research of the last forty years has reached back already into the deep recesses of the second millenium and is giving us an increasingly full and accurate picture of life in that remote period. The rapid and widespread industrial development of the last decade, involving extensive excavation, has brought to light so great a mass of artifacts of ancient Chinese civilization that it is estimated that at least twenty years will be required for their adequate examination, classification and interpretation.

In face of this great wealth of historical material, all that is attempted in the following pages is to select a few topics to demonstrate the influence which geography has had upon them. This will give a geographical background to some of the outstanding developments in Chinese history.

CHINESE ORIGINS

'The Neolithic age, disdained by pre-historians because it is too young, neglected by historians because its phases cannot be exactly dated, was nevertheless a critical age and one of solemn importance among all epochs of the past, for in it Civilization was born . . . In a matter of ten or twenty thousand years man divided up the earth and struck his roots in it.'
                    (Pierre Teilhard de Chardin, *The Phenomenon of Man*)

Whence has sprung the great Chinese people, who today number nearly one quarter of the people of the earth? Who are they? Did their great and continuous civilization develop *in situ* or does it owe most of it to importa-

45

tion and invasion from surrounding lands? If it is indigenous, what natural factors have influenced that development? These are some of the questions which have engaged the attention of Chinese scholars through the centuries and Western scholars in more recent years.

Until quite recently, when modern Western scholarship introduced the concept that earliest civilization had its birth in a common centre in the Near East from which it has radiated, the Chinese accepted as valid their classical mythology, which attributed their beginnings to legendary heroes, beginning with the creator, P'an Ku, and the pastoral age under Fu Hsi in 2953 BC. These myths fell into disrepute and gave place to the Western theory that nomadic tribes of Turko extraction spread out eastward from Turkestan and gradually, in the course of many generations, migrated into the Tarim Basin and Dzungaria, across the southern borders of the Gobi, and, following the grass covered slopes of the Nan Shan, crossed the Hwangho, descended into the basin of the Wei Ho, a land flowing with milk and honey, and there settled to become the Chinese people. It was not until the third decade of this century that this theory, in its turn, was seriously challenged. The questioning arose from the growth and spread of archaeological survey and research in north China. The work of a group of devoted and eminent pioneers, including J. G. Andersson, Davidson Black, Pierre Teilhard de Chardin, C. C. Young, V. K. King, W. H. Wong and C. D. Wu, has revealed evidence which suggests that Chinese origins are to a large extent indigenous; which evidence largely, though not entirely, discredits both Chinese mythology and Western migration theories.

In 1921 and 1926 the remains of very early man, *Sinanthropus pekinensis* —Peking Man, were found in a cave at Chou Kou Tien to the south of Peking, giving proof of the presence of very primitive man in China some 500,000 years ago. He was probably later than *Pithicanthropus erectus* and earlier than Neanderthal Man. It is contended that the jaw and skull present affinities with those of the modern Mongol.[1] Then comes a long gap in our knowledge. The next trace of human beings in this area was found at Chou Tong K'ou, about 8 miles east of Ninghsia, where remains of Palaeolithic Man, probably about 50,000 years old, were found. Andersson suggests that he lived there at the end of the wet Pleistocene period and at the beginning of the dry period which followed when the thick deposits of loess were laid down.[2] This was the period equivalent to the Würm glaciation in Europe when, in China, the bitter, dry wind carrying the

[1] F. Weidenreich, '*Sinanthropus* Population of Choukoutien', *Bulletin of the Geological Society of China*, 1935.

[2] J. G. Andersson, *Children of the Yellow Earth* (London 1934), p. 142.

loess were inimical to both vegetation and animal life.[1] Again a long blank period follows and the trace of man is not picked up until about 3,000 BC, when evidence of Neolithic Man is found widely and thinly spread over China.

At Yang Shao Tsun, not far from modern Loyang, extensive finds were made in 1922 by Andersson of neolithic artifacts—polished stone tools, axes, adzes and hoes, bone ornaments and fine painted pottery, great round-bottomed burial urns and domestic pottery with bold and flowing designs in red and black. Further finds of a similar but not identical character were made in the upper Hwang-ho in Kansu at Ma Chia Yao, at Pan Shan and in the Koko Nor area at Ma Ch'ang. This pottery, known as Yang Shao, is not unlike neolithic pottery of both western and central Asia, and points to there having been considerable intercommunication at that early time (3,000 BC). In the last decade many more such sites have been discovered in north China, emphasizing how widely Neolithic Man was spread. Nevertheless, Yang Shao pottery has several unique shapes, which re-appear in subsequent cultures of north China and are found nowhere else. These support the belief that Chinese civilization rises from an indigenous culture. The unique shapes are the *li*, a very efficient tripod cooking vessel with hollow legs, shaped rather like goat's udders, sometimes large enough to cook, say, a goose in one leg, an antelope in another and vegetables in a third.[2] A second unique shape, the *hsien*, a kind of sieve or steamer, is fitted above the *li*, the two together forming the *hsien*.

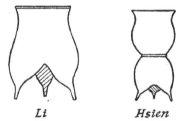

Li   *Hsien*

FIG. 18. Characteristic Yang Shao pottery-shapes

Yang Shao Man lived on the loess plateau at a time when the water-table must have been much higher than it is today, and before the countryside was cut up by deep ravines, as this section (after Andersson) shows:

[1] W. Watson, *China* (London 1961), p. 28.

[2] It is interesting to note the development of the character *li* which today means large earthenware pot or iron cauldron: 鬲 鬲 鬲 鬲

While he certainly was a hunter, as his arrowheads and scrapers indicate, his large limestone hoes and adzes also point to his practising primitive agriculture. 'In one coarse, thick fragment [of pottery] we observed imprints, which have been shown to be husks of rice (*Oryza sativa L.*), which was thus cultivated at Yang Shao Tsun by the people of the late Stone Age.'[1]

In 1929 further discoveries were made, this time at Cheng Tzu Yai, east of Tsinan in Shantung. The finds were of Black Pottery, hard and highly polished, wheel-made and of very fine texture. This culture is known as the Lung Shan. It continues the shapes of the Yang Shao, notably the *li* and the *hsien* and apparently is unconnected with anything found in the West. Wherever Black Pottery is found in conjunction with Yang Shao Pottery, it is found lying above it, indicating that it is later in origin. The fauna revealed by the finds at Yang Shao Tsun consisted of only pigs, dogs and oxen. To these the sheep and the horse are added in the Black Pottery finds.[2]

In 1934–5 extensive and careful excavations were carried out near Anyang. This work led to exciting discoveries. The site of the ancient capital city of the Shang or Yin dynasty was uncovered close to the modern Anyang, revealing a highly organized society and an advanced civilization. Here was a firmly settled people in a planned town, sited cleverly in the bend of a river. Quite elaborate houses with pounded earth walls and floors and gable roofs at least for the ruling class had replaced the neolithic bee-hive shaped huts. While stone and bone tools, implements and weapons were still in general use, elaborate and exquisitely shaped bronze vessels, mainly for sacrificial and ceremonial purposes, are found in the tombs of royalty and the aristocracy. Significantly, some of these bronzes are cast in the same unique shapes that we noted in the Yang Shao and Black Pottery. The tombs, too, yield evidence of clothing, of handicrafts, of painting and music. Writing in the form of pictographs on oracle bones had made its appearance by this time.

[1] J. G. Andersson, op. cit., p. 180.
[2] Li Chi, *The Beginnings of Chinese Civilization* (Washington 1957).

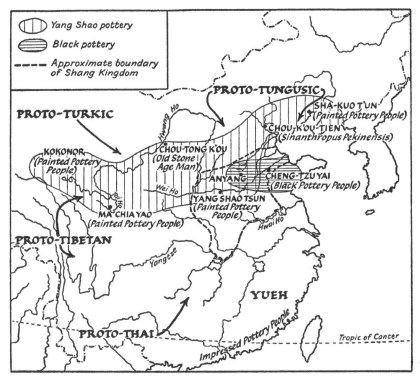

FIG. 19. Sites of Painted Pottery and Black Pottery cultures

We must pause at this point for some explanation of oracle bones, which are important in throwing light on the origins of the Chinese people. They are, in fact, the main source book of history before 1122 BC. Country folk of north China have, for centuries, found pieces of bone on which they noticed curious markings. These were termed 'dragon bones' and were sold to the medicine shops as being specially efficacious. A big sale of these in the late nineteenth century led Chinese scholars to take note of them. Early in this century these markings were recognized as very early writing, but it was not until later that scholars, notably Tung Tso Ping, were able to decipher them. Then, and only then, were the oracle bones recognized for what they were and their true importance realized.

The bones were scapula (shoulder blades) or leg bones of cattle, or carapaces of the tortoise. Prayers or advice asked of the gods were incised on them. Unincised bones had been used in Black Pottery times, but by mid-Shang writing had appeared. Oval holes were cut on the surface of the bone and a hot iron placed to one side of these holes. This resulted in

49

a crack appearing on the other side of the bone. The cracks were always along the length of the hole and then to one side 〈 𝆑 𝆑 , the simple answer 'yes' or 'no' being divined from this angle. The system by which this was done has not been worked out. The questions were on all sorts of everyday subjects, very largely of an agricultural nature, asking, for example: 'Will there be enough rain for the crops?' 'Will the harvest be good?' or making a request such as: 'We pray for rain.' Wheat and millet figure frequently, and hemp and even rice are mentioned. This mention of rice might indicate that in Shang times the climate was wetter than today, and the constant concern about rain an indication that desiccation was taking place. What is of significance is the almost exclusive concern in these oracle bone inscriptions for agriculture: for the crops, for rain, for the harvest. Sheep are rarely mentioned, indicating that we are dealing with a settled agricultural people. Just as the Hebrew imagery of the Old Testament (e.g. The Twenty-third Psalm) is cast in a pastoral mould, so that of the Shang is essentially agricultural, as one or two examples will show:[1]

> 'If, the father having broken up the ground, his son is unwilling to sow the seed, how much less will he be willing to reap the grain.' 'Heaven in destroying Yin is doing husbandman's work—how dare I but complete the business of my fields.'[2]

In the early stages of development of any society, environment exercises its influence much more powerfully than it does in later stages. We are justified in looking for the development of a settled stable society from a primitive nomadic society only in very favourable environments in which its peculiar advantages can be put to specialized use.

There are good geographical reasons for believing that a differentiation in economy into pastoralists and agriculturalists in north China was taking place in neolithic times. The region at the confluence of the Hwang Ho and the Wei Ho provided an environment in which this could occur. The rivers themselves and some of their tributaries in this area were less liable to flood than lower down the course. Here was a light, fertile loess soil, which could be easily worked with the crude stone implements of Neolithic Man. The natural vegetation was mainly grassland, thus eliminating the heavy work of forest clearing, although probably there was also considerable woodland since heavier rainfall was experienced then than now. The presence of some

[1] G. H. Creel, *The Birth of China* (Baltimore 1937).
[2] J. Legge, *The Chinese Classics* (London 1861–72), Vol. 3, pp. 371–2.

woodland would enable the primitive agriculturalist to supplement his food supply by hunting—the way of life from which he was emerging. This, too, is an area where variability of rainfall is beginning to be marked but not serious, and this factor would tend to make man thoughtful and forward-looking, a necessary quality in the development of civilization. From this the art of irrigation and water control eventually developed and with it the co-operation and organization that it implies. Neolithic man availed himself, as does the modern Chinese in this area, of the peculiar vertical cleavage and ease of working of the loess for housing. Caves, which are dry and cool in summer and warm in winter, can be cut with comparative ease. These are some of the geographical factors which led to early settlement in this region.

Farther north and north-west climatic conditions were too hostile for agricultural development; rainfall is sparse (less than 15 in. per annum), the seasonal range of temperature increases and the growing period becomes very short. To the east the lower reaches of the Hwang Ho were subject to such constant flooding that they were unusable in the state of knowledge of those times. The Yangtze valley to the south and west at that time was a region of dense forest and marshland, occupied by thinly distributed aboriginal tribes, relying mainly on hunting and migratory *ladang* agriculture.

As we shall see later when discussing Chinese growth outward from its primary focus in the Wei-Hwang Ho bend, while expansion was comparatively easy, rapid and permanent southward and eastward, no such movement was possible north-westward. To the south and east the same agricultural economy as that of the focal centre, with modifications occasioned by local differences, could be established. Dry agriculture had to give place to wet agriculture in the south, but agriculture it remained. To the north and north-west this was not so. The increasing aridity north-westward dictates, at least to a primitive culture, a pastoral way of life. Attempted agricultural expansion north-westwards met with increasingly adverse conditions. More and more reliance had to be placed on rearing livestock on the steppe grassland until a region is reached where dependence on grazing is complete and a true pastoral economy is established.

There thus grew up two ways of life, which were mutually exclusive and which were hostile to each other: the Chinese (Shang) retained their hunting habits but developed more and more their agricultural activities, while the Mongols placed more and more reliance on their sheep, goats, camels and horses. Lattimore suggests that the two ways of life did not become finally exclusive until the 'barbarians' of the north developed horse riding and the use of a bow on horseback, while the Chinese continued to use the horse harnessed to chariots.[1] Between the true agriculturalist (Chinese) and the true pastoralist (Mongol) there has lain a transitional zone, which throughout history has been contended for and which is marked geographically by the 15 in. isohyet and historically by the Great Wall.

THE GREAT WALL AND ITS FUNCTIONS

'Obviously a line of cleavage existed somewhere between the territories and peoples that could advantageously be included in the Chinese Empire and those that could not. This was the line that the Great Wall was intended to define.'

Owen Lattimore[2]

We have seen how, out of the Lung Shan culture, there emerged a highly organized people based on the Great City *Shang* or *Yin*, having a king or *Ti* and an elaborate hierarchy of classes. However, they were not yet fully sinicized. For example, theirs was largely a matriachial system. This, the first dynasty (Shang 1450–1054 BC), was eventually conquered by a small kingdom, Chou, itself essentially nomadic, under the leadership of Wu Wang, in alliance with many nomadic tribes of the west. The Chou people were probably of Turkish stock and occupied the land lying to the west of Shang. The Chou introduced a patriarchial system and the worship of Heaven by the Emperor, and ancestral worship, all of which played so important a role in Chinese history from then until the fall of the Manchu dynasty and the rise of the Republic in 1911.

When the Chou defeated and overran Shang lands, the only way of governing this newly acquired and comparatively vast area was to divide it up and distribute it as fiefs to the lords and tribal leaders, who had been instrumental in achieving victory. Thus it came about that China entered a long feudal period (1054–256 BC). Chou territory proper, i.e. the royal domain, was sited in two areas, the original home in the west, having Hao (Sian) as its capital and a newly acquired realm in the heart of the kingdom, with Loyang as its centre.

The Chou were never a very big tribe themselves; nevertheless the first emperors of the Chou dynasty exerted a strong central control. However,

[1] O. Lattimore, *Inner Asian Frontiers* (London 1940), Chapter 3.

[2] O. Lattimore, 'Origins of the Great Wall', *Geographical Review*, Vol. 27, 1937, p. 546.

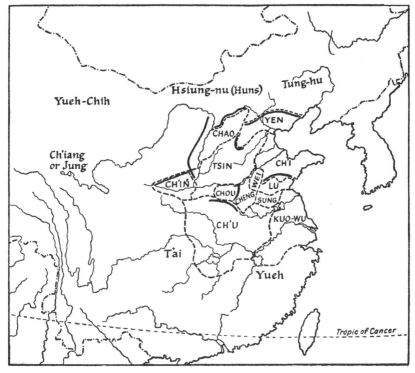

FIG. 20. Inter-state walls *c.* 350 BC

as time went on, the various fiefs and dukedoms became virtually independent kingdoms, vying with the centre and each other for power. After establishing themselves, the Chou were constantly engaged in subduing and controlling the Yueh Chih, Mongols (Hsiung-nu) and Tibetans (Ch'iang) on their northern and western borders. Eventually their western domain was overrun and the Chou prince had to flee to Loyang. The territory was regained by the prince of Ch'in, who retained it for himself. Thus the Chou prince became merely the nominal head but continued to carry out the sacrifices which were considered so essential for securing harmony between heaven and earth, without which there could be no general well-being.

From now on (*circa* 770 BC) there was a constant struggle between the feudal princes and lords for power. This is the period of the 'Warring States' when Confucius and Mencius expounded their theories of good government and the right relationship of heaven to prince, and prince to subject. The farther the fief was from the centre, the less it was subject

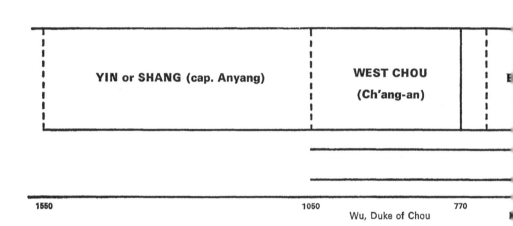

YIN or SHANG (cap. Anyang)

WEST CHOU

(Ch'ang-an)

E

1550   1050   770

Wu, Duke of Chou

IRON BEGINN

DISUNITY   *UNITY*   D

INVASION FROM NORTH & WEST   GREAT EXPANSION

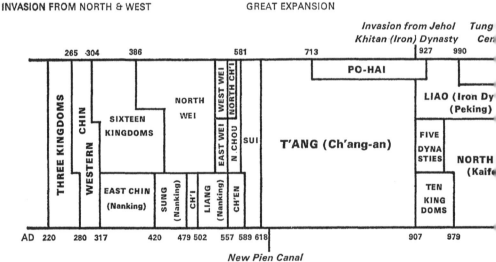

Invasion from Jehol   Tung
Khitan (Iron) Dynasty   Cer
927   990

265 ·304   386   581   713

PO-HAI

THREE KINGDOMS

WESTERN CHIN

SIXTEEN KINGDOMS

NORTH WEI

WEST WEI

NORTH CH'I

EAST WEI

N.CHOU

CH'EN

SUI

T'ANG (Ch'ang-an)

LIAO (Iron Dy
(Peking)

FIVE DYNA STIES

NORTH
(Kaif

EAST CHIN

(Nanking)

SUNG (Nanking)

CH'I

LIANG (Nanking)

TEN KING DOMS

AD 220   280 317   420   479 502   557 589 618   907   979

New Pien Canal

KEY ECONOMIC AREA = HONEI   *KEY ECONOMIC AREA = YANGTZE*

FIG. 2

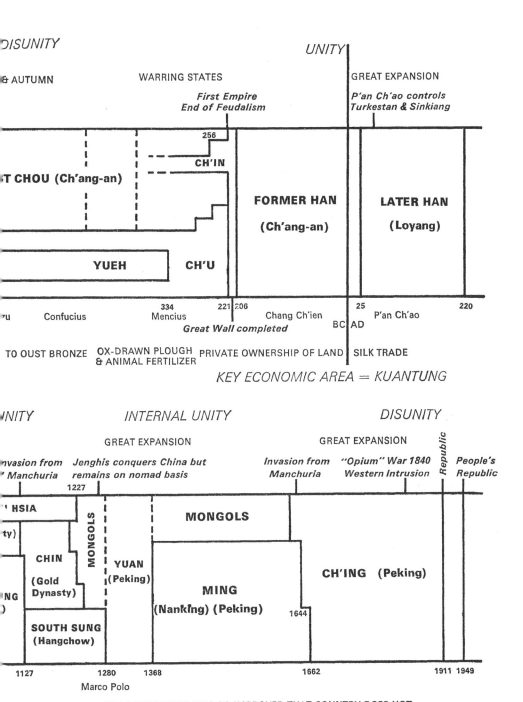

Historical time chart

to control. Moreover, the nearer it was to the 'frontier' and the steppe-lands, the more it was subject to nomadic influence. There was a persistent danger that those lords holding lands on the northern periphery would throw off their Chinese allegiance and culture and go over to a pastoral way of life, thus adding weight to the constant steppeland pressure on the Chinese intensive agricultural way of life, centred on the walled city.

We have seen how, in 771 BC, the feudal prince of Ch'in rescued the Chou emperor in the west and took over his domain there. In 256 BC the Chou dynasty came to an end, abdicating to a later head of this same state of Ch'in. By 222 BC the young Ch'in prince, Ch'in Shih Hwang Ti, had subdued all the other feudal states. He then proceeded to create the first unitary state of China. He centralized control in every sphere. He set up military and civil governors in all districts, making them directly responsible to the emperor. He attempted to formulate a universal language and writing. He standardized the length of cart axles—a very necessary measure as the varying lengths rendered the dirt roads unusable. He nearly succeeded in destroying all Confucian literature, since Confucius was the great upholder of the feudal tradition that the Emperor was attempting to destroy.[1]

Ch'in Shih Hwang Ti further tried to unify the state by stabilizing its northern frontiers. As we have seen, this frontier posed an eternal problem, not only of resistance to invasion from the surrounding tribes but also of loss of the frontiersman to the pastoralist way of life. It was to solve this problem that Ch'in Shih Hwang Ti built the Great Wall. In feudal times, the various princes had built many walls along their borders to secure them against the inroads of their neighbours, both Chinese and barbarian (see Fig. 20). Ch'in Shih Hwang Ti, by means of immense forced labour, untold misery, blood and sweat, set to work and connected all the individual walls of the feudal princes on the north and extended them right round the great northward bend of the Hwang Ho as far as Ninghsia. Later, under the Han emperors, this wall was extended far into the west to Tun Huang and Yumen (Jade Gate) to protect the Imperial Silk Route against the Hsiung Nu (Huns) of the north. Ch'in Shih Hwang Ti thus occupied and attempted to hold the whole of the Ordos by settling some 30,000 Chinese families in this semi-desert region. However, the attempt was doomed to failure. Within a century the Hsiung-nu, who had been driven out, were back in full occupation of the Ordos.

From Shanhaikwan on the shores of the Gulf of Liao-tung to the Hwang Ho, Ch'in Shih Hwang Ti's Great Wall followed the highlands of the southern rim of the Mongolian basin and thus had some physical

[1] See Leonard Cottrell, *The Tiger of Ch'in* (London 1962).

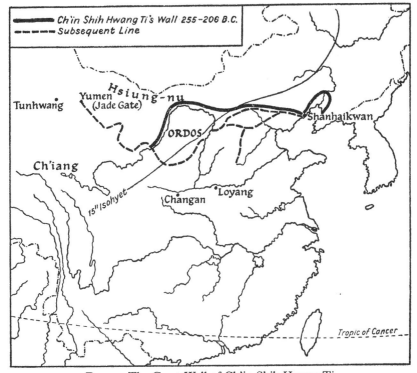

FIG. 22. The Great Wall of Ch'in Shih Hwang Ti

justification. However, in its continuation westward along the north bank of the Hwang Ho the Wall ceases to conform to a natural region, for it crosses the 15 in. isohyet and embraces a large area of sparse and variable rainfall, the Ordos, which is far more suited to pastoral economy than intensive agriculture. Thus, in disregarding geographical factors and attempting to include permanently within his domains essentially pastoral lands, Ch'in Shih Hwang Ti defeated his own ends and the main purpose of the Wall, i.e. the separation of these two economies.

Often there were large numbers of nomads living within the Great Wall while it was sited so far north. Nineteen Hsiung-nu tribes occupied all the Ordos region at the time of the Three Kingdoms (AD 220–265). While the Han emperors remained powerful and energetic, they were able to keep the northern pastoralists under control, but immediately there was a weakening of imperial power, the old forces reasserted themselves and the struggle between the two ways of life was renewed. Ch'in Shih Hwang Ti's wall to the north of the Ordos was eventually abandoned, and one to the south, conforming closely to the 15 in. isohyet, was built.

For more than two millennia there has been an ebb and flow across a zone of transition between these two ways of life. It has fluctuated as first the agriculturalist and then the pastoralist has had an accession of power and an urge to energetic action. Particularly in recent centuries, as population in China has rapidly increased, there has been an outward pressure by the agriculturalist to colonize the more arid regions beyond the Wall, but until today, the 15 in. isohyet has been the inexorable controller, and the Chinese farmer, on the whole, has remained within the confines of the Wall.

Now, once again, a strong central government in China is pushing its culture far beyond the Great Wall and the 15 in. isohyet into the lands of the northern pastoralist, imposing its own intensive methods of agriculture on the steppe-land people. This modern attempt, however, differs from all such movements in the past in that it is today fortified by all the techniques of modern science as applied to irrigation and power, which may be able to defy the old dictation of climate.

CHINESE EXPANSION AND DEVELOPMENT

We have seen how the natural advantages offered by the loess region enabled Neolithic Man to develop a primitive specialized agriculture from general undifferentiated neolithic culture. Later, early Chinese under Shang and Chou leadership developed intensive agriculture and were able to establish stable government and a settled community in the valleys of the lower Wei and middle Hwang Ho, a district known as Kuan-chung. Ssu-ma Chien, writing of it, claimed that it 'occupied one third of the territory under heaven with a population of three-tenths of the total; but its wealth constitutes six-tenths of all the wealth under heaven'.[1] We have seen also that the extension of agricultural economy and way of life towards the north and north-west was fraught with great difficulty. The high Tibetan plateau in the west held out no invitation. If expansion was to take place, therefore, it had to be to the east and the south.

It was to the east, out over the alluvial plain of the lower Hwang Ho, into the region known as Honei, that Kuan-chung first naturally expanded. As Pan Ku, extolling the natural strength and advantages of Kuan-chung, put it, 'the abundance of Shu and Pa (Szechwan) on the south, the nomadic herds of the Hu on the north and, with natural barriers on three sides, it can easily be defended; with one side opening to the east it is an excellent

[1]Ssu-ma Chien, *Historical Records* (*Book on Rivers and Canals*), Chuan 29.

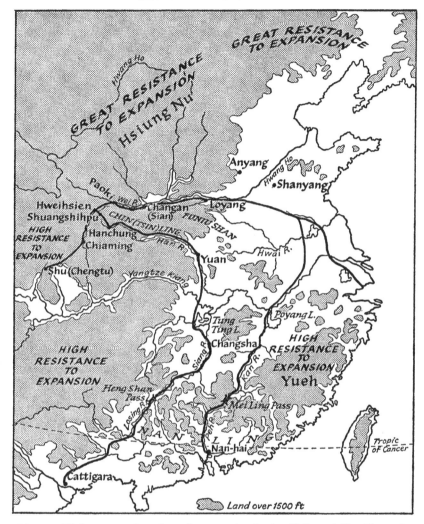

FIG. 23. Main routes of southward penetration in the Ch'in and Han Dynasties

base for the subjugation of the feudal lords'.[1] By the end of the Chou dynasty the whole of the North China Plain as far north as the walls of Yen and Chao (present Shansi, Hopei and Shantung) was occupied by Chinese princedoms and kingdoms. The Ch'in and Han dynasties, which followed, marked a period of great development and change politically, socially and economically. China was united under one ruler and govern-

[1] Pan Ku, *Earlier Han History*, Chuan 40 (*The Biography of Chang Liang*), p. 8.

ment; feudalism was abolished and the peasant was no longer bound to the soil. It entered the Iron Age, the plough and animal fertilizer were introduced and great water conservancy works were carried out.

Routes south

It was not until the great conquests under Ch'in Shih Hwang Ti and the Former Han that any considerable expansion south of the Tsinling was made. Penetration southward followed five comparatively easy topographical lines.

The easiest way was that south-eastward over the almost imperceptible divide between the lower Hwang Ho and the Hwai basin. This route, more restricted in width then than now on account of marshland and poor drainage, was destined to become of great strategic importance once the Yangtze basin had been developed and incorporated in the Middle Kingdom. It formed a bottleneck between north and south. Writing in the Sung period (AD 960–1280), when threatened by invasion from the north, Shu Tsung-yen says 'Shanyang is a place over which north and south must fight. If we hold it, we can advance to capture Shantung, but if the enemy get it, the south of the Hwai can be lost in the next morning or evening.' This Hwai route forks farther south, the eastern prong leading to the mouth of the Yangtse and to Hangchow, and the western prong up the Yangtze to the Poyang Lake, thence southward via the Kan River, over the Mei Ling Pass and down the Pei Kiang to Nan-hai (Canton) at the mouth of the Pearl River. East of this route is the very dissected, rugged highland of south-east China. This is the country which the Yueh tribes occupied and which resisted the arms of both the Ch'in and the Han.

A second route southward had two starting points, Chang-an and Loyang. That from Ch'ang-an crossed the Tsinling by the Wu-ling Pass and descended to the Han valley at Yuan. It was here that the route from Loyang joined it after crossing the Fu-niu Shan. From Yuan it went south via the Tung Ting Lake and the Siang River to Ch'ang-sha, over Heng Shan and Nan Ling, forking east to Nan-hai (Canton) and west to Cattigara (Tongking).

A third route was more difficult. It ran west from Ch'ang-an to Paoki, thence south over the high passes and very difficult terrain of the Tsinling at Shuangshihpu. There it divided, one branch going eastward down to the source waters of the Han at Han-chung and continuing, joined the second route, described above, at Yuan. The main route, however, at that time branched south-west via Hweihsien and Chia Ming to Shu (present day Chengtu) and so into the heart of the State of Shu-Oa (Szechwan). This was the route followed by Ch'in Shih Hwang Ti when he subdued Shu in 220 BC and also by Kublai Khan about AD 1280 as a means of outflanking the Middle Yangtze.

Reasons for Southern Expansion

Movement southward along one or other of these routes was occasioned, as may be expected, by many causes. Under Ch'in Shih Hwang Ti and the early Han emperors there was a rapid and vast expansion of imperial territory south, extending to the Pearl River and to Tongking. This expansion was due in large measure to the perennial ambition and the desire for prestige of the emperors themselves and their generals. However, one must not be misled by the maps into imagining that this vast new area was very much under central control or that the land was closely settled by immigrants from the north; in fact it was held for centuries only by widely dispersed garrisons, often giving a very doubtful loyalty.

Population pressures appear to have had little to do with southward movement if the following figures are any guide:[1]

Period	Year AD	Population	
Western Han	2	59,595,000	
Eastern Han	156	56,487,000	
T'ang	755	52,919,000	probably an under-estimate in order to avoid taxation
Northern Sung	1102	43,822,000	
Yuan (Mongol)	1209	59,847,000	
Ming	1578	60,693,000	
Ch'ing (Manchu)	1783	284,033,000	probably an over-estimate in order to flatter the court. Also tribes not before included.
Ch'ing (Manchu)	1851	432,140,000	
Communist	1951	563,000,000	
Communist	1954	582,603,417	First reliable census.

The figures show a more or less static population over nearly 1,600 years. Nevertheless there have been periods of mass migration southward caused by invasion from the north-west by the Hsiung Nu or Mongols and, to a

[1] H. J. Wiens, *China's March towards the Tropics* (Hampden, U.S.A. 1954), p. 169.

61

lesser extent, when natural disasters of flood or drought have occurred. Wiens quotes four such occasions, which are known as 'panics' and each has been the result of invasion and internal disorder. Between AD 299 and 317 the Western Ch'in dynasty broke up as a result of intrigue and misrule, coupled with a series of droughts. Subordinate princes in alliance with strong frontier tribes, notably Hsiung-Nu, invaded the country, which resulted in mass migration of the Han Chinese from Kuanchung and Honei southward into Szechwan, Hupeh and Hunan.[1] A Ch'in prince, Yuan Ti, formed the Eastern Ch'in dynasty in the south with its capital at Nanking. 'Countless members of the Chinese gentry had fled from the Huns at that time and had come into the southern empire. They had not done so out of loyalty to the Chinese dynasty, or out of national feeling, but because they saw little prospect of attaining rank and influence at the courts of the alien rulers, and because it was to be feared that the aliens would turn the fields into pasturage.'[2]

There have been three other such 'panics'; one when the Toba (Turks) conquered the north at the end of the eleventh century; another when Jenghis Khan and his Tartars swept into northern China at the beginning of the thirteenth century; and lastly when the Japanese invaded China in 1937, followed shortly after, in 1949, by the sweeping victory of the Chinese Communist armies.

A further reason for migration southward and subsequent colonization was the promise which the fertile lowlands of the Yangtze basin held out. Although a great deal of it was heavily forest clad, the potentialities of the Yangtze valley as a granary were not lost on the early settlers, and it became, in due course, the economic heart of the Middle Kingdom. Wiens comments:

'No doubt the psychological appeal of the southern rice regions has had its influence in effecting a continuous migration southward. Since the earliest days the south has been regarded as a granary for the north, and in this manner, the north early became a parasite upon the south. First the grain of Ssu-ch'uan was drawn upon to support northern political and military power. This became especially significant after the construction of the Grand Canal.'[3]

When the Han Chinese moved into the Yangtze valley and Szechwan, they found it far from unoccupied. Eberhard estimates that of the 800 or so tribes and folk groups in China at this time, 345 were on the south-west

[1] W. Eberhard, *A History of China* (London 1952), pp. 119–26.
[2] ibid., p. 162.
[3] H. J. Wiens, op. cit., pp. 170–1.

and west, 290 in south China and only 80 in north China. In the west and south-west outstanding tribes were the Ch'iang (62), Lolo (93) and Fan (32), while in the centre and south were the Miao (65), Pai Man (44) and Yao (32). Many of these tribes were known collectively as the T'ai people and formed the state of Ch'u, which long resisted intrusion from the north until conquered by Ch'in Shih Hwang Ti in 222 BC. Wiens says:

'The political fall of the T'ai in the Yangtze valley, Szechwan, south-eastern China and Lingnan occurred under the onslaught of the other mighty barbarian state, pressing upon the crumbling Chou Empire from the north-west. Fortunate it was for the Ch'in that the various semi-sinicized T'ai states were not united, for such a comity of states would certainly have thwarted his ambitious schemes to seize supreme power. All of south China at this time was dominated by the T'ai. In fact, as Eickstedt points out, most historians talk about a "China" existing prior to the victory of the Ch'in that not only never existed but was actually a T'ai empire (Eickstedt, pp. 115–30). This, he rightly asserts, ought to be recognised once and for all.'[1]

While the Miao, Yao, Lolo and many other groups of tribes were hill rovers, existing on gathering and hunting almost entirely, the T'ai were the primitive agriculturalists of the south, who, using milpa or ladang methods, practised wet rice cultivation. 'The custom was to fertilize the land by burning the vegetable overgrowth and, as the seeds were planted, flood the fields with water.'[2] It was from the south, too, that the Shang and Chou dynasties obtained the copper and tin for their magnificent bronze work, and it is contended that they also learned their earliest techniques from the south. The *Tribute of Yu* in the Shu Ching written in the Chou dynasty, lists also gold and silver, feathers, hair, ivory, hides, woods, silken fabrics, pearls, large tortoises and cinnabar as coming from the southern regions.[3]

Ssu Ma Chien[1] describes Ch'u and Yueh, i.e. Lower Yangtze, as 'a large territory, sparsely populated, where people eat rice and drink fish soup; where land is tilled with fire and hoed with water; where people collect fruits and shellfish for food; where people enjoy self-sufficiency without commerce. The place is fertile and suffers no famine and hunger. Hence the people are lazy and poor and do not bother to accumulate wealth.

---

[1] H. J. Wiens, op. cit., pp. 128–9.

[2] Wei Ching, 'History of Sui Dynasty' (*Book of Food and Commodities*), Chuan 24, p. 3.

[3] James Legge, *The Chinese Classics* (London 1861–72), Vol. 3, Part 1, p. 92.

Ssu Ma Chien *Biographies of Merchants and Industrialists* Chuan 102, p. 12, quoted from Ch'ao-ting Ch'i, *Key Economic Areas in Chinese History*.

Hence in the south of the Yangtze and the Hwai, there are neither hungry nor frozen people, nor a family which owns a thousand gold.'

The southern penetration under Ch'in Shih Hwang Ti met with strenuous resistance. Although the T'ai were defeated in the Yangtze valley, they continued to hold Yunnan and Kweichow. This region remained in T'ai hands as the Kun-chou or Nan-chou Kingdom until about AD 1000 The Yueh also resisted strongly, and, retreating into the forested hills of the south-east, were not subdued by either the Ch'in or Han dynasties. Szechwan, known then as Shu-pa, proved a hard nut for the early expansionists from the north to crack—and not only for these but for many succeeding invaders, not excluding the Japanese in the Sino-Japanese war of 1937–45, who never succeeded in cracking it. Liang Ch'i-Ch'ao says of Szechwan 'Whenever there were disturbances under heaven [i.e. in China], Szechwan was held by an independent ruler, and it was always the last to lose its independence.'[1] The physique of Szechwan with its very difficult mountainous borders on all sides is sufficient explanation of this phenomenon.

The method of colonization used by the early conquerors as they pushed south, was that of planting the invading armies as agricultural settlers under the leadership of their generals, who were appointed as governors. In spite of the fact that Ch'in Shih Hwang Ti attempted to plant half a million such settlers they were thinly distributed over a large area, and through intermarriage with the native population, they were strongly influenced by the 'barbarian' culture. As has so often been the pattern in history, the generals or leaders who have been instrumental in conquering new lands and who, as a reward, have been given control of the peripheral areas, soon asserted their independence. For example, Chao T'o, Ch'in Shih Hwang Ti's most successful general, proclaimed himself King of Nan Yueh, although he continued to give nominal allegiance to the central power. He took to himself a Yueh wife and to some degree identified himself with the Yueh culture.

Defecting generals and chieftains were loath to sever entirely their ties with imperial authority as they valued very highly their identification with the 'civilized' culture of the north as against the 'barbarian'. We should remind ourselves that the distinction between 'civilized' and 'barbarian' in Chinese eyes, until quite recent years, has been 'our' culture as against all others. China did not call herself 'the Central Kingdom' for nothing. The high value set on maintaining this connection with the northern culture is reflected in the fact that the king or prince would be willing to concede

[1] Liang Ch'i-ch'ao, 'Essay on Chinese Geography' *Yiu Pin Shih Collected Essays* (Shanghai 1926).

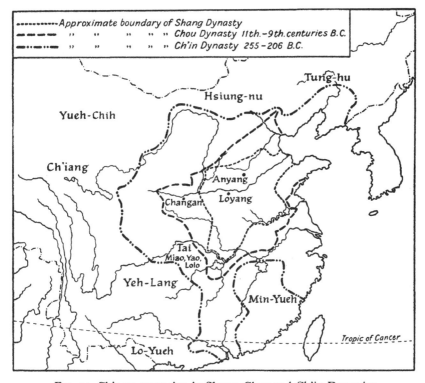

FIG. 24. Chinese expansion in Shang, Chou and Ch'in Dynasties

vassalage to the emperor. This willingness played an important part in imperial expansion. The great prestige of Han culture and the desire to be included in its aurora was certainly as potent a factor as its military prowess in extension of its imperial rule.

However, real conquest of the south did not come until T'ang and Sung times. To this day the people of China living south of the Nan Ling in Kwangsi and Kwangtung refer to themselves as T'ang Jen (Men of T'ang) while those to the north and centre call themselves Han Jen (Men of Han). While the maps (Figs. 24 and 25a) of the Ch'in and Han periods show large areas of present day China as under central sway, we have seen that much of it was held nominally only and was by no means fully sinicized. This was particularly true of all that area south of the Nan-Ling, known as Lingnan (Kwangsi and Kwangtung) and, of course, Yueh (Fukien and Chekiang), which did not give even nominal allegiance to the north. With the fall of Han and later the Three Kingdoms, connection with imperial power either at Ch'ang-an (Sian) and Loyang was even more tenuous and

65

for long periods was entirely severed, for China was divided into hostile states and was not reunited until the Sui dynasty.

There were geographical reasons for the slow assimilation of Lingnan into the imperial fold. The hot, wet valleys of Kwangtung and Kwangsi were malaria-infested and certainly held little attraction for the Han Jen from the cooler, drier north. Curiously, the area of the upper reaches of the West River in Kiangsi apparently were settled earlier than the lower reaches and wider valley in Kwangtung. This was probably due to the more frequented use of the Siang River—Lotsing River route, leading to Cattigara (Tongking) than that of the Kan River—Peh River, leading down to the mouth of the West River at Nan-hai (Canton). The resurgence of imperial

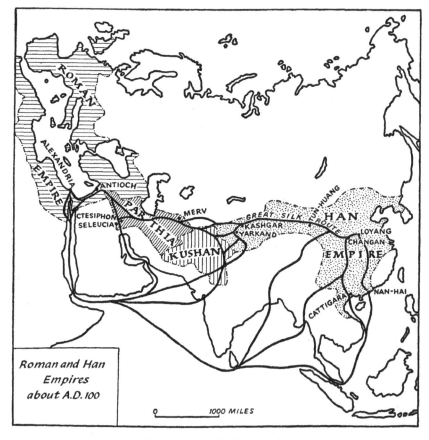

FIG. 25a. Imperial boundaries

power and expansion which followed the reunification of the country under the glorious T'ang dynasty (AD 618–906) led once more to the active planting of more military farm colonies in the south, and when, after the fall of T'ang, the succeeding Sung emperors (AD 960–1279) were forced south from their capitals in Ch'ang-an, Loyang and Kaifeng and established their headquarters at Nanking and Hangchow, the whole of the south was fully occupied and completely sinicized. So it has remained ever since an integral part of 'China Proper' inhabited by Han or T'ang Jen as distinct from those wide alien areas which have been incorporated into the Middle Kingdom at times of imperial exuberance and ambition, and which have fallen away as soon as that energy has expended itself.

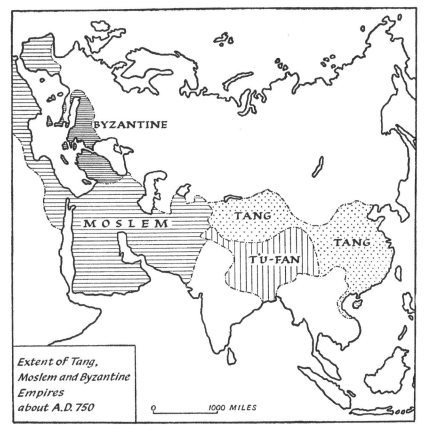

FIG. 25b. Imperial boundaries

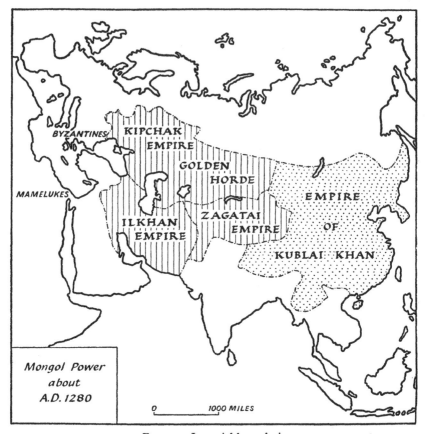

FIG. 25c. Imperial boundaries

HISTORICAL DEVELOPMENT OF WATER CONSERVANCY AND COMMUNICATIONS

'When water benefits are developed there will be good results in agriculture, and when there are good results in agriculture the State's treasury will be enriched.'

Gazetteer of the Prefecture of Suchow

The Legend of Yu has it that there was a Golden Age in China between 2357 and 2205 BC. The Emperor Yao reigned from 2357 to 2255 and during this time he called on his minister Yu to cope with the floods, which were

68

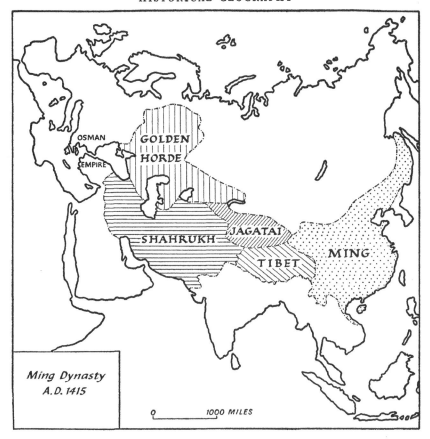

FIG. 25d. Imperial boundaries

devastating the country. This Yu did with great energy, fidelity and self abnegation.

'Yu checked the overflowing waters. During thirteen years, whenever he passed before his home, not once did he cross the threshold. He used a chariot for land transport; a boat for water transport; a kind of basket for crossing marshes and crampons for crossing the mountains. He followed the mountains in separating the nine provinces; he deepened the river beds; he fixed tribute according to soil capacity; he made the nine roads usable; he dammed the nine marshes and surveyed the nine mountains.'[1]

To cope with the flood he used many methods. He is said to have divided

[1] *Book of Hsia* (See E. Chavannes, *Les Mémoires Historiques de Ssu Ma Ch'ien* (Paris 1898), Chapter 29, 17th Treatise: 'Canals and Rivers').

Greatest Extent
of Manchu Empire
1760–1842

0 _____ 1000 MILES

FIG. 25e. Imperial boundaries

the lower course of the Hwang Ho into nine channels, a method which, as we shall see, was carried out so successfully by Li Ping in the Cheng-tu region in Szechwan. Already, at this early time, the raised bed of the Hwang Ho (referred to as 'The Ho') was causing difficulty.

'Then Yu, taking account of the fact that the Ho's bed was raised, that its waters were rapid and violent and that in crossing the plain they caused much damage, he drew off two canals in order to control its course.

In the north he carried the Ho on raised terrain beyond the R. Kiang to Tu-lou. He divided it into nine rivers, which he re-united to form the Ni-ho and emptied into the Po-ho.

When the nine courses had been well cut and the nine marshes cleaned [i.e. drained], the whole empire was orderly and at peace. This meritorious work was beneficial throughout three dynasties.'[1]

[1] E. Chavannes, op. cit.

Yu further used the obvious but much less satisfactory method of containment by building dykes along the river banks, a method continuously practised to the present day and equally continuously the source of disaster following the inevitable periodical breaking of the banks.

Ssu Ma Ch'ien records one such breakage in the early days (168 BC) of the Han Dynasty and another some thirty-six years later:

'The Ho overflowed at Hou-tzu, spread out over the swamps of Chiu-yi and linked up with the Hwai and Szu. Then the Son of Heaven [i.e. the Emperor] sent Chi Yen and Ch'eng Tseng-chi to recruit men to close the breach but it suddenly broke through again.'[1]

The river then flowed out south of the Shantung Peninsula. There is an interesting little comment in the same record showing that care and repair of dykes was not always pursued with enthusiasm or viewed with an altruistic eye.

'At that time T'ien Fen, Marquis of Wu-nan, was Prime Minister and was in receipt of the revenues of the city of Chow, which lay to the north of the Ho. After the breach in the Ho, the waters flowed south; the city no longer suffered from flooding and its revenues increased. Therefore T'ien Fen addressed the Emperor in these terms: "The breaches of the Kiang and the Ho are of heaven's ordaining; trusting to human power it is not easy to close them; moreover, it is by no means certain that to do so is the will of heaven." The necromancers gave the same advice and so, for a long time, the Son of Heaven made no effort to close the breach.'[2]

However, after a lapse of twenty years, during which time, as a result of the breach, there had been a series of bad harvests, capped by a severe drought, the Son of Heaven was forced to take some action. So, in 110 BC we find him issuing orders for the mobilization of a vast labour force of several hundred thousand workers to close the breach. He visited the site in person where he propitiated the River God by throwing into the river a white horse and a jade ring and, having performed the necessary rites and sacrifices,

'He ordered all his subjects and officials, from the grade of general down, to carry faggots to fill the breach. At that time the brushwood had been burnt off and small wood was consequently scarce. Bamboo from the Ch'i woodlands was therefore cut and used for barrage stakes. The Son of Heaven, approaching the breach of the River, was grieved that the work was not finished and sang this song:

[1] E. Chavannes, op. cit.
[2] ibid.

There is a breach at Hou-tzu—what should be done?
It is an inundation, an immensity—there, where hamlets stood, there is
naught but the River.
Since there is naught but River, the country can enjoy no peace.
This is not the time to relax effort—our mountains are breaking down.
Our mountains are breaking down and the Kin-yi swamp overflows.
The fish are agitated and ill at ease, tortured by the approach of winter.
The whole length of the River's bed is damaged: the River has abandoned
its regular course.
Alligators and dragons dart forth: they wander afar in complete liberty.
When the River returns to its ancient bed, 'twill be the blessing of the God.
How should I have known what was happening outside the capital had I
not performed the *feng* and *chan* sacrifices?
'Tell me, Lord of the River, why are you so hostile?
Your inundations cease not and you desolate my people.
Yi-sang is submerged; the Hwai and Szu overflow.
Long have you left your bed: the rules governing the waters are spurned'."

Evidently either the sacrifices or the hundreds of thousands of workers
were successful, for a further verse runs:

The bubbling waters rush along their courses.
Turning towards the north, they return to their bed.
Alert, they flow over all obstacles.
Take up long poles: cast on the beautiful jade.
The Lord of the River is answering our prayers.
But the wood is insufficient. Why? The fault lies with the men of Wei.
Fire has laid the land waste.
Alas, how shall we stop the waters?
Cut down the bamboo forest: drive in the stakes and place the stones.
Ten thousand congratulations—at Suen-fang the barrage is secured.

The Hwang Ho was thus turned northward once more and resumed its
old course, traced by the legendary Yu.

While the purpose of dyke building was primarily to prevent flooding,
it was sometimes used as a weapon in the quarrelsome feudal days of the
Chou dynasty (the fifth and fourth centuries BC). Feudal lords would
construct dykes in such a way as to direct flood waters into their neighbour's
country with the express purpose of embarrassing them.

'Those who were anxious to avoid the danger of floods also constructed
dykes to force water into their neighbour's country, regarding the latter as a
reservoir for surplus water. Thus more and more dykes were built day by
day and they encroached so much upon the natural channel of the river that
the dykes were burst and floods became frequent.'[1]

---

[1] Cheng Hsiao, *Words about Ancient Things*—quoted from Ch'ao-ting Ch'i,
Key Economic Areas in Chinese History (London 1936).

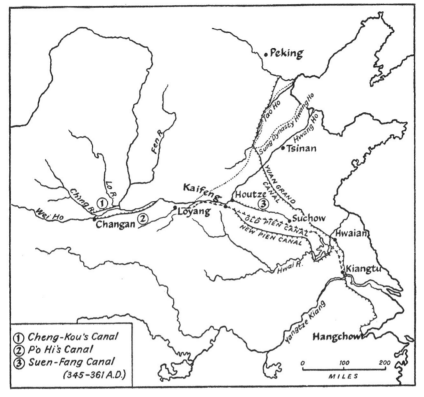

FIG. 26. Canal development

Ssu Ma Ch'ien tells a typical Chinese story of an attempt by the head of the State of Han to use canal building to weaken its western rival, the State of Ch'in. This attempt was singularly unsuccessful: it recoiled on its instigator and helped materially towards the ultimate conquest by Ch'in of all the feudal states, and the formation of the first unitary government of China.

'Then the Prince of Han, learning of the success of the state of Ch'in, wished to exhaust it and to oppose its attacks to the east. He therefore sent a hydrographic engineer, named Ch'eng Kou to the Prince of Ch'in and treacherously persuaded him to undertake the construction of a canal, which would lead the waters of the River King from Ch'ung Shan west to Hou-kou along the whole length of mountains to the north, emptying into the River Lo. The canal would be 300 *li* long and would be used for irrigation. The work was half finished when the ruse was discovered. When Ch'in would have killed Ch'eng, the latter said: "True, at first I deceived you; however, if the canal is finished, it will be of great benefit to Ch'in". The Prince of Ch'in agreed

73

and ordered the completion of the canal. When finished it irrigated 40,000 *ch'ing* of alkaline land and production rose by one *ch'ong* per *mou*. Thus, the interior became a fertile plain without bad years and Ch'in became rich and powerful conquering other feudal lords. This is how the canal came to be named The Ch'eng Kou Canal.'[1]

However, throughout the Chou dynasty, i.e. until 221 BC, water conservancy was on a relatively small scale. The whole economy was feudal and based on the well system, by which each square *li* of land was divided into nine squares each of 100 *mou*. The central square was a public field, the produce of which belonged to the lord. The other eight squares were occupied and cultivated by eight peasant families, who were bound to their holdings, and who paid a tenth part of their produce as tax. Moreover, work on their lord's square had to be completed before they could turn to their own. As long as this feudal system held sway no large scale water conservancy work could be done since no large labour force, bound as it was to its feudal farm routine, could be organized. The breakdown of this system came when first the shortlived Ch'in dynasty (221–206 BC) broke the power of the feudal lords and then the institution of private ownership of land under the Han emperors freed the peasant from his feudal bondage. At much the same time great technical changes in the use of the land were taking place. The use of iron was replacing bronze and stone, oxen were being used for ploughing and animal fertilizer was applied to increase the productivity of the land.

One of the effects of this agricultural and social revolution was to set free an army of peasants and serfs, who could be gathered into a labour force such as we have seen used in dealing with the breach in the Hwang Ho. This ability to mobilize large numbers was the necessary condition for ushering in a great era of canal building and irrigation, two aspects of water control which went hand in hand and which were essential to the development and the expansion of China outlined in the previous chapter. Irrigation was necessary if the production of grain and foodstuffs was to keep pace with the growing population and the ever-increasing demands of the court and central government. Canals and navigable rivers were necessary for the transport of tribute grain to the capital, 'road' transport being hopelessly uneconomic. Once central government by the Emperor was established under the Ch'in and Han dynasties, tribute grain became of great importance and retained its importance until well into the nineteenth century. The functions of tribute grain were threefold: it formed the main supply for the maintenance of the emperor and his court in the capital,

[1] E. Chavannes, op. cit., Chapter 29, 17th Treatise.

which was synonymous with the government; furthermore, it maintained the army on which the government relied to carry out expansion of its territories, to defend it against invasion and possible rebellion; and part of the tribute grain was stored in the Imperial granaries against times of famine. The efficient functioning of these granaries was a mark of able and stable government.

We have seen how Ch'eng Kuo's Canal came to be built as a result of a ruse by the Prince of Han in the feudal times of Chou. In Ch'in and Han times this canal continued to be of local use in the heart of Kuanchung, transporting grain of the Wei valley to the capital, Ch'ang-an, and irrigating the alkaline land lying between the left bank tributaries of Ch'ing and Lo. However, with the extension of government jurisdiction farther and farther eastward into the lower Hwang Ho district of Honei, the need for easier communication and transport to the capital along the line of the Hwang Ho and Wei became more pressing. Ssu Ma-chien records the building of a further canal in these words:

'At that time [approx. 132 BC] Ch'eng Tang-che was Minister of Agriculture. He observed "Formerly transport of grain came by the passes up the course of the River Wei and required six months for its accomplishment. The route by water was about 900 *li* and sometimes was very difficult. If the waters of the Wei were led by a canal, which could be cut from Ch'ang-an, skirting the mountains on the south, to reach the Ho, the passage would be a mere 300 *li* and direct. Transport would be easy and a matter of only three months. Moreover, more than ten thousand *k'ing* of cultivable land below the canal could thus be irrigated. By this means, on the one hand water transport would be shortened and the numbers engaged reduced and on the other hand, the fertility of the land above the passes would be increased and greater harvests obtained." The Son of Heaven approved this project. He ordered the hydraulic engineer, Sui Po, native of Ts'i, to plan the canal and to recruit all available manpower to carry it out. The work was finished in three years and proved most beneficial. As a result, water transport gradually increased and the people who lived below the canal often used its waters for irrigation.

'Later P'o Hi, administrator of Ho-tung, said: "The amount of grain that is transported by water each year from east of the mountains to the west is about a million *che*. Passage of Ti-chou has lost much of its difficulty, fatigue and expense".'[1]

A great deal of irrigation work was carried out in Kuanchung and Honei under the early Han emperors. The great Emperor Wu Ti, in 111 BC was

E. Chavannes, op. cit., Chapter 29, 17th Treatise.

most energetic. He issued an edict laying the duty of irrigation on his government in these terms:

'Agriculture is the basic (occupation) of the world. Springs and rivers make possible the cultivation of the five grains . . . There are numerous mountains and rivers in the domain, with whose use the ordinary people are not acquainted. Hence (the government) must cut canals and ditches, drain the rivers and build dykes and water tanks to prevent drought.'[1]

Big irrigation works were carried out in Ninghsia and Suiyuan. Ch'eng Kuo's canal was extended and supplemented by subsidiary canals between the Rivers Ch'ing and Lo. However, some of the work lower down the river was rendered useless by one of the Hwang Ho's periodic changes of course. P'o Hi comments:

'When the canals carrying the waters of the Fen for irrigation of the land below P'i-che and Fen-yin and the waters of the Ho for the land below P'ou-fan are pierced, I estimate that five thousand *k'ing* of cultivable land will have been gained: land which until now has been only waste land beside the Ho, where people cut their hay and tended their herds. Now, if they irrigate the land, I reckon they will be able to reap more than two million *che* of grain. This grain will go up the River Wei and will be similar to that of the interior of the passes. Undoubtedly, one would not provide transport for grain coming from east of Ti-cheu. The Son of Heaven approved this project and great numbers of workers were recruited to make the canals and fields. After several years the Ho changed its course and the canals were of no use; the farmers reaped no harvest. In the end the canals and fields to the east of the Ho were abandoned. The land was given to the men of Yu and the *chao-fu* was ordered to be gentle in his tax collection.'[2]

At about the same time as these big irrigation and transport schemes were being carried through by Emperor Wu Ti, the first governor of Cheng-tu (Shu), Li Ping, planned and executed one of the world's most remarkable hydraulic engineering achievements. The plain of Cheng-tu, measuring approximately 70 miles from north-east to south-west and 40 miles from north-west to south-east, was at that time an old lake bed, with stony desert in the north and rank marshland in the south. It was the recipient of many mountain torrents issuing from the steep edge of the Azure Wall range, which marks the western border of the Red Basin. These mountain torrents spread over the area in constantly changing courses. Li Ping began to convert this unpromising region into cultivable land. The work was completed by his son, Li Erh-lang.

[1] Ch'ao-ting Ch'i, op. cit.

[2] E. Chavannes, op. cit.

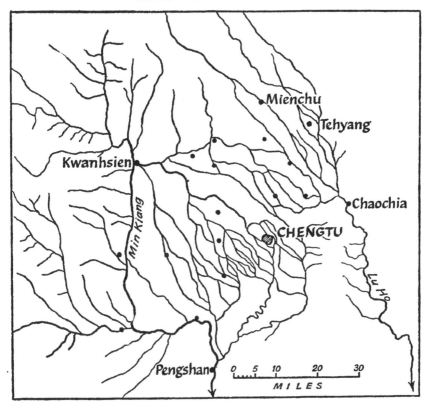

FIG. 27. Cheng-tu irrigation system

The main river descending to the Cheng-tu plain is the Min. It cuts through the Azure Wall mountains at Kwan-hsien. In summer it is a torrent, flowing over a bed half a mile wide. Even in winter, when the water is at its lowest and quietest, it flows in a stream 50 yards wide and 6 feet deep. Acting on the principle of divide and rule, Li Ping split the river at Kwan-hsien, leaving one half to flow southward in its old bed and turning the other in a new course eastward. He then proceeded to subdivide these two main streams into innumerable channels, covering the whole area with a net-work, which irrigated and indeed still irrigates the plain, turning it into probably the most densely populated, the most fertile and the most pro-ductive rural area in the world. In order to effect these divisions, Li Ping built 'arrow heads' of faced stone in mid-stream and supported these and the banks with enormous 'sausages', 30 ft. long and 2 ft. in diameter, of boulders, bound together in a network made of bamboo strips. This method is still used today, having been used continuously for more than twenty centuries. Moreover, Li Ping coined a maxim, a guiding rule, which is cut in the granite of the gorge at Kuanhsien and which reads: 'Shen t'ao t'an, ti tso yen' (深 淘 灘 低 作 堰) meaning 'Dig the beds deep; keep the dykes low.' This rule the farmers of Szechwan have religiously observed through the centuries to their profit and so have avoided the disasters which have attended the dyke building of the north.

There were no major additions to the work of the Li's, father and son, for more than one thousand years. During Yuan (Mongol) times the channels were increased, the dykes lined with faced stone, cemented with lime and wood oil (t'ung yu) and the banks planted profusely with willows.[1]

For more than 350 years after the fall of the Later Han dynasty China was a disunited country. At first it was divided into the Three Kingdoms, which occupied three natural regions: Wei in the north occupied the middle Yellow River area; Shu in the west held the Red Basin and Wu in the east held the middle and lower Yangtze. These broke up and were replaced successively by the West and East Chin, the Sixteen Kingdoms and the Wei Kingdoms. The Wei period (AD 386–535) was one of considerable Buddhist and Indian influence religious, social and political. The Lattimores write:

'Its monastic communities were important in advancing the techniques of a collective economy. Although they did not have the family type of heredity, from father to son, they did have corporate continuity. They made possible the pooling of individual knowledge and skill; they held large tracts of land; their farming was prosperous and progressive, and they carried on the great Chinese engineering techniques of irrigation, drainage, the prevention of floods and the building of transport canals.'[2]

In spite of the turbulence and disunity of this long period, quite a lot of water conservancy work, some of it of a complex character as in the Hwai valley, was carried out. However, it was localized and on a comparatively small scale.

The country was once again brought under one government when the Sui swept in from the north-west in AD 581. Both the first emperor, Sui Wen-ti and the second, Sui Tang-ti, set about the work of unification with vigour. The problem was to draw together the political centre of gravity

---

[1] A. Little, *The Far East* (London 1905), Chapter 6, gives a fuller account of Li Ping's work.

[2] O. and E. Lattimore, *The Making of Modern China* (London 1945), p. 87.

PLATE I: The Great Wall, which divided the Ordos from the loess, has also laid down a rough line of demarcation between the Mongolian pastoral and the Chinese arable way of life.

PLATE II: A remarkable example of hydraulic engineering was the division of the River Min by Li Ping in the second century B.C. Basically the same method of reinforcing the banks and 'arrow heads' with boulders is still used today.

in the north and the economic centre of gravity, which was shifting rapidly to the rice lands of the Yangtze valley. To do this it was essential that good water communication be established between the granary in the south and the capitals of Ch'ang-an and Loyang in the north.

There already existed a fragmentary old route. In the north Ch'eng Kuo's and P'o Hi's canals, running from east to west, linked Honei with Kuantung. Loyang was linked with the Hwai valley by a canal (the Old Pien Canal) which ran via Kaifeng, along the Ssu River via Hsuchou to Hwai-an. A further canal ran south from Hwai-an to Kiang-tu (present day Yangchow) and so to the Yangtze. However, neither of these two last-named canals was in good repair or of adequate size. Accordingly, Sui Yang-ti embarked on the ambitious plan of cutting a new waterway, known variously as the New Pien, the T'ung Ch'i or the Grand Canal, from Kaifeng to Hwai-an. It ran south of the Old Pien, following close to the Kwei and Hwai Rivers so to Hwai-an. The section from Hwai-an to Kiang-tu, which had been both shallow and narrow, was greatly improved, being widened to forty paces and tree-lined throughout. The whole canal from Ch'ang-an to Kiang-tu was liberally equipped with post stations and imperial resting places. The work was carried through, as were all such large scale projects, by dictatorial forced-labour methods. It is recorded that more than five million men were pressed into service and that there was great loss of life owing to the great hardships and also the cruelty of the administrators. The resulting unpopularity did not a little to bring about the fall of the Sui dynasty. Sui Yang-ti 'shortened the life of his dynasty by a number of years but benefited posterity to ten thousand generations. He ruled without benevolence but his rule is to be credited with enduring accomplishments.'[1]

During the T'ang dynasty (AD 618–907) the Yangtze valley became established as the key economic region of the Empire and has remained so until the present time. This New Pien Canal from Sui and T'ang times until the Mongol conquest in AD 1279 was of vital political and economic importance in holding together north and south.

Their vast conquests made it inevitable that the Mongol conquerors of 1279 should be greatly concerned with communications. Kublai Khan chose to regard China rather in the nature of a separate dominion, but still he could not afford to ignore the remainder of his great empire. Therefore, in order to be in touch with both China and the lands to the north and west, the capital was moved north from Loyang to Peking. Thus the

---

[1] Yu Shen-hsing. Quoted from J. Needham, *Science and Civilization in China* (Cambridge 1954), Vol. 1, p. 123.

political centre was even farther removed from the key economic area of the Yangtze. For this reason the route of the Grand Canal was changed. Instead of turning west at Hwai-an in the northward journey, it followed the Ssu valley as far as Hsuchou, and then, turning north, picked up an old discarded (Sui Dyn.) bed of the Hwang Ho, thence to Peking. This enormous engineering feat was again carried out by means of massed forced labour with its usual accompanying cruelty, hardship and discontents.

With the fall of the Yuan and the rise of the Ming dynasty (1368) the capital was moved for a short while to Nanking but was very soon transferred back to Peking. There it remained until the fall of the Manchus in 1911. The need, throughout this long period was the same as during the time of Kublai Khan, i.e. the maintenance of good communications between the political north and economic south. The story, too, was the same in both dynasties—an energetic beginning during which time the Grand Canal was repaired and improved and then, as the dynasty waned and became decadent, it fell into disrepair and largely into disuse.

CHINA'S TRADE ROUTES

From very early times Imperial China has regarded trade with the barbarian without its walls with a certain superiority and aloofness. It has held that China produces all that is necessary for good living within its own borders and its rulers have said so in no uncertain terms on more than one occasion. For example, Emperor Ch'ien Lung in an edict addressed to George III on the occasion of Lord Macartney's embassy to Peking in 1794 said . . . 'nor do we have the slightest need of your country's manufactures'. Even in China itself trade and the merchant were graded low in the social scale. The scholar-landlord-administrator was ever vigilant to see that the trader was kept in his place and excluded from governmental office.

Nevertheless, the emperor and his court were not averse from receiving what the outside world had to offer and a good deal of interprovincial and foreign trade was carried on under the guise of 'tribute'. The 'barbarian' envoy, paying court to the emperor, presented his tribute and in return was laden with gifts, which were at least to some extent a *quid pro quo*. Far more, however, passed through the hands of merchants.

There is a description in the *Book of Records* of what is known as 'The Tribute of Yu' (*Yu Kang*), being the movement of goods from the Nine *chou* or provinces to the Imperial Residence at Anyang about 1125 BC.

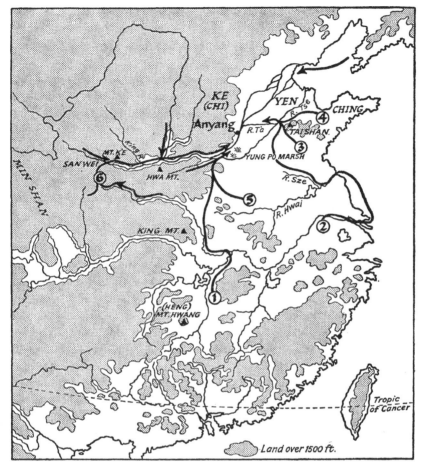

FIG. 28. The Tribute of Yu: 1. copper, gold, silver, ivory, hides, timber, feathers, cinnabar; 2. copper, gold, silver, bamboo, precious stones; 3. silk, pearls, fish, feathers, varnish; 4. salt, timber, silk, hemp, precious stones, varnish; 5. silk, hemp, tung yu; 6. furs, wild animals, silver, precious stones

The very varied produce which travelled to the capital is shown on the accompanying map.

In early times (Han dynasty) by far the most important and desired commodity which China had to offer was silk. This was in very great demand in imperial Rome; so much so that, it is said, a pound of silk was worth a pound of gold. Curiously, it was not the finely woven Chinese brocades and damasks that were wanted in Rome, for when they arrived they were unravelled and re-woven into lighter, flimsier silk gauzes. Other

Chinese exports were cinnamon and such medicinal roots as rhubarb. Rome paid for these largely in gold and silver but also in dyes, woollen textiles and glass, which, before China learned to make it herself, was looked on as precious as jade.

The Chinese have never been an outstandingly maritime people. The rugged south-east coast has produced some redoubtable sailors—and pirates—but on the whole the Chinese have looked landward rather than seaward.

Land Routes

The Silk Route. China's east-west communications have been very largely dictated by physical geography.

Between China and Burma lie heavily folded mountains, giving rise to a series of deep valleys and high ranges, running north and south, which are particularly difficult to cross, the more so as the hot, damp river valleys are malaria-infested and the mountain sides are heavily forested. Thus, although the natural barriers are not impassible, this region has not lent itself to easy communication.

Tibet lies to the west. The high mountain ranges which form its eastern borders lead to a vast plateau, barren of anything that would attract the merchant.

There is, however, a route to the north which leads along the northern slopes of the Nan Shan, through the Jade Gate (Yumen) to the Tarim Basin, over the Pamirs at Kashgar to Afghanistan and Turkestan, thence to Persia and down into the Tigris-Euphrates basin and so to the Mediterranean. It was along this route that most of the merchandise between China and Rome in Han times flowed. From Ch'ang-an to Rome was a matter of some 7,000 miles. When it is remembered that the only means of transport available were camel, pack mule and horse, yaks, sheep and human beings, it will be realized that only luxury goods of very high value in proportion to their bulk would pass along this route. Thus it was that silk was the principal commodity going westward, and gold, silver, glass, amber and precious stones eastward.

In Han times there was no direct contact along this route between Rome and the Seres, as the Chinese were called by the Romans. It may be that very occasionally a Roman merchant made the whole journey, but it was not until Marco Polo that we find any account of such an achievement. Trade passed between merchant and merchant through the various stages of the route and these stages were determined by the changing political authority

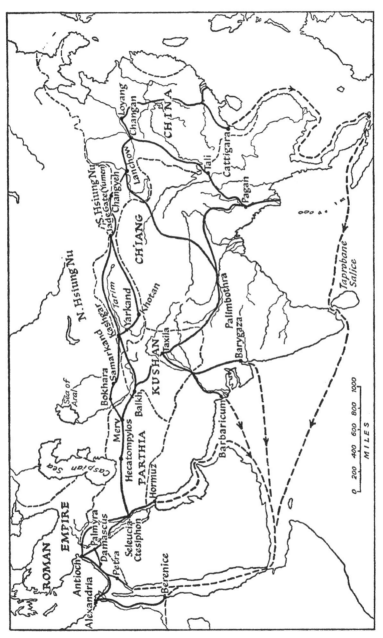

FIG. 29. East-West communications AD 100

exercised in the region, which, in its turn, was greatly influenced by physical geography.

The route may be divided into four main stages[1] and the first stage ran from Ch'ang-an to Kashgar. Starting from the capital, Ch'ang-an or Loyang, it crosses the Hwang-ho at Lanchow and utilizing the narrow but generous strip of steppeland lying between the high ranges of the Nan Shan on the south and the Gobi on the north, reaches Yumen (Jade Gate). The Great Wall was extended as far as Changyeh (Kanchow) in order to protect the northern flank of the 'road' against the Hsiung-nu (Huns). This was a very vulnerable part of the route owing to the presence of the Hsiung-nu on the north and the Ch'iang (Tibetans) on the south and was under Chinese control only when there was a strong central government, as under the Han, T'ang, Yuan and Ch'ing (Manchu) dynasties.

Passing west of the Jade Gate, the traveller meets one of the most difficult sections of this stage. The road divides north and south as it begins to skirt the eastern end of the Taklamakan. Both roads have to traverse 150 miles of the most hostile desert. It is practically rainless, has few depressions or wells and no grazing for camels. The Imperial Silk Road is the southern of these two roads. Chinese travellers and merchants, more often than not, were obliged to use this southern and more difficult route because the Hami region was usually a Hsiung-nu centre. Both the southern road, which follows the foot of the Astin Tagh (Altyn Tagh), and the northern road, which keeps to the southern slopes of the Tien Shan, are true routes as distinct from nomadic lines of march.[2] They are trade routes, linking oasis with oasis in true desert. In the nomad occupied semi-desert the needs of the flocks and herds are paramount, and trade is subsidiary. Movement is over a wide 'fairway'. The two north and south trade routes round the Taklamakan desert meet again at Kashgar at the western end of the Tarim basin, whence the road crosses the Pamirs. However, before following the Silk Route to its second stage we must note two southern branches from this route, both leading to India.

One route, little used but interesting because its line is followed approximately by the newly built motor road to Lhasa, runs through the Ts'aidam and then south across the Tibet plateau to the Brahmaputra, and so on to the Ganges plain at Palibothra (Patna), thence by sea westward to Rome. The second branch left the Silk Route at Khotan and entered India either

[1] G. F. Hudson, *China and Europe* (London 1931), gives a full treatment.

[2] O. Lattimore, 'Caravan Routes in Inner Asia', *The Geographical Journal*, Vol. 72, 1928, pp. 497–531.

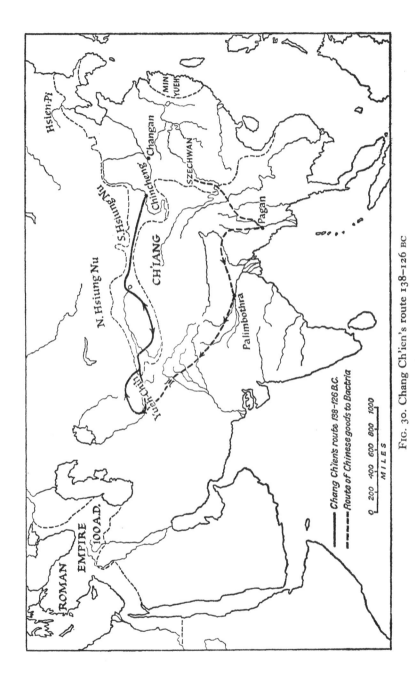

Fig. 30. Chang Ch'ien's route 138–126 BC

85

by the Karakoram Pass or Khyber Pass and so to the coast either at Barbaricon or Baryzaga. This was the road so well trodden by Buddhists making their way to and from China to India.

The second stage carried the road from Kashgar over the Pamirs, down the valley of the Alai, into the lowland of Bactria (Turkestan), through Samarkand to Merv. This region first came into Chinese ken in the reign of the great Han emperor, Wu Ti. The Hsiung-nu were constantly harrying the northern Han border and had also defeated another large nomad group, the Yueh Chih and driven them away westward. Wu Ti conceived the idea of searching out this tribe and supporting them against the Hsiungnu. Accordingly he sent an envoy, Chang Ch'ien, in 140 BC to find this tribe and form an alliance. The misfortunes and adventures of Chang Ch'ien in his search make a classic Chinese epic. He eventually found the Yueh Chih in Bactria, where they had defeated the Greeks and settled down comfortably. They showed no disposition to renew their struggle with the Huns. Chang Ch'ien was surprised to find some Chinese goods—cloth and bamboo—already in Bactria and discovered that they had come via India. The significance of Chang Ch'ien's mission is that it eventually led to two Chinese military expeditions against the King of Ferghana in Bactria, the second of which was successful and led to considerable increase in trade and intercommunication.

The Persian plateau between Merv and Seleucia formed the third stage. Both this and the second stage had the advantage of remaining in the hands of one ruler over longer periods than either the first or fourth stages and therefore enjoyed more security for trade. The fourth stage from Seleucia to the Mediterranean coast went either to Palmyra or Damascus, mainly along 'the Way', notorious for contenders for its use and control.

Serious interruption in communication along this east-west line came with the decline and fall of the Han dynasty. China lost control of the Tarim basin; silk became more difficult to obtain in the west and prices soared. It is not surprising, therefore, that the secret of sericulture, religiously guarded by China for so long, was lost. Eggs, moth and knowledge of silk processes were smuggled out. Various tales are told of how this was done: a Chinese princess was betrothed to a King of Ferghana and brought them with her in her trousseau; Buddhist monks on pilgrimage smuggled them out; merchants brought them in the hollow of bamboos. Be that as it may, sericulture began to flourish first in Syria, then in Italy and Spain and by the end of the sixth century AD Europe was virtually self-sufficient in silk. As a result trade, contacts and interest in the Far East decreased.

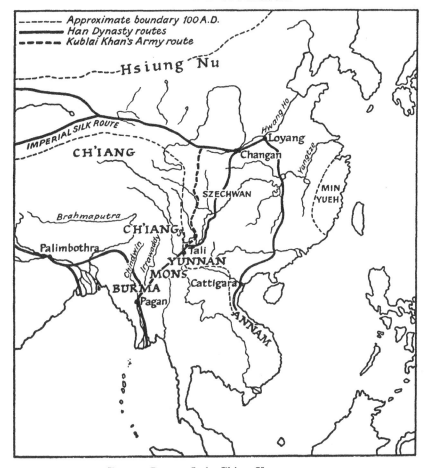

FIG. 31. Burma, Indo-China, Yunnan routes

European interest was rekindled in the seventh and eighth centuries when Christendom looked hopefully but unsuccessfully to the powerful China of the T'ang dynasty as an ally on the eastern flank against militant Islam. Again later, for a short period there was considerable east-west overland trade during the Mongol regime (AD 1280–1368) when, in addition to the Silk Route, the nomadic lines of march across the Gobi and Dzungaria, farther north, were much used.

Burma, Indo-China and Yunnan. There are no formidable physical barriers between China and Indo-China. Early Chinese expansion under Ch'in Shih Hwang Ti reached out into Annam (214 BC) before penetrating into Lingnan, i.e. Kwangtung. Even the high parallel ranges to the west, though very difficult, are not impassable. The Mons, Khmers and Shan, remnants of which tribes are still in the highland borders, probably originated in south-west China and moved over into Burma as Chinese pressure from the north increased.

We have already seen how Chang Ch'ien noted that Szechwanese goods were being sold in Bactria and that they had come via Burma and India. On his return home he urged Emperor Wu Ti to develop this route, and this the emperor attempted to do. Silk and medicinal herbs moved up the Yangtze from Szechwan to Tali on the Yunnan plateau, thence down the Irrawaddy to Pagan, which became the trade centre. From Pagan, goods moved either down-river to Pegu or up the Chindwin and across the Naga Hills into Assam. Two Buddhist priests are said to have reached China via the Irrawaddy and Yunnan between AD 58 and 75.[1]

The Mons constantly blocked this route. A campaign was led against them in the Three Kingdoms period (*circa* AD 230). The Mons chieftain was captured seven times, released and told to come back with his men, until at last the Mons gratefully laid down their arms and remained true to their promise never to block the route. However, later the Tibetans, who subsequently occupied the region, were not so co-operative and we find the Governor of Szechwan, Chang Chien-chih, petitioning the Emperor in AD 698 in these terms:[2]

'Yunnan was kept open in earlier dynasties because through Yunnan China was connected in the West with Ta-tsin (e.g. India or the Roman Empire) and in the south with Indo-China. Tax receipts in cloth and salt are now declining and precious tributes fail to come. We press our people to garrison the territories of the tribes to no purpose. I therefore suggest that we withdraw from Yaochow (midway between Kunming and Tali) and leave the tribes south of the Lu River (Yangtze) as vassal states and prohibit all traffic except by special permission',

and the road was officially closed by the Emperor.

After the fall of T'ang the Turks and Tartars blocked the Imperial Silk Route in the north and for a while the route was used unofficially. It was not until Kublai Khan sent his armies into Burma that it was opened officially again and then only for a matter of forty to fifty years. It is interesting to note that Kublai Khan's army came south via Sikang to Yali

[1] Kuo Ts'ung-fei, 'A Brief History of the Trade Routes between Burma, Indo-China and Yunnan', *T'ien Hsia*, Vol. 12, No. 1, 1941.
[2] ibid.

and not through Szechwan and used much the same route as that used by the Red Army in their Long March northward in 1934. With the fall of the Yuan dynasty the route again fell into disuse and was not revived until the Second World War, when the Burma Road was constructed.

Sea Routes

Westerners did not sail the China Seas via Malaya until after AD 47 when Hippalus discovered the use of the monsoons. The Arabs had a virtual monopoly of sailing between the Red Sea and Malaya, and the Chinese a monopoly east of Malay. China was quite a considerable maritime power in Han times. A large fleet of 2,000 naval junks was used to suppress a revolt in Indo-China and secure control of that area. Sea traffic was increased when the Imperial Silk Route was virtually blocked at the fall of the Han dynasty and the rise of Parthian power. Much of the silk trade was carried all the way from China to Ta Ts'in (Arabia Felix) and Roman Egypt by sea, although probably never in the same bottoms all the way. Transhipment took place probably at Taprobane Salice (Ceylon).

First contacts with Japan were probably made after the conquest of Manchuria and Korea by Wu Ti (109–108 BC). *The History of the Later Han* records the first Japanese envoy to China in AD 57. By T'ang times there was a considerable sea trade with Japan and Korea. Entry into China was by way of either the mouth of the Hwai or Yangtze at Kiang-tu or modern Hangchow, thence by canal to the capital, Ch'ang-an. No use was made of the Hwang Ho, which was unnavigable. The normal crossing from Japan to China took from five to ten days. There was a considerable traffic of Buddhist monks and pilgrims. Incense and medicines were carried from China to Japan.[1] In addition to this Japanese trade there was considerable Arab and Persian trade in T'ang times. Kiang-tu (Hangchow) was the chief port. Some idea of the volume of trade and the importance of Kiang-tu can be gathered from the report that several thousand Arab and Persian traders were killed there during anti-foreign riots in AD 760.

We have seen how the transfer of the imperial capital from Ch'ang-an to Peking in Yuan (Mongol) times drove Kublai Khan to build a new Grand Canal from the granary of the Yangtze to the north. This transference had its influence also on Mongol shipping. A speedy and safe sea route for the transport of grain was needed. Two former salt smugglers

---

[1] E. O. Reischauer, 'Notes on T'ang Dynasty Sea Routes', *Harvard Journal of Asiatic Studies*, Vol. 5, 1940–1, pp. 142–64.

and pirates of the Yangtze delta, Chu Chin and Chang Hsuan, were given the office of superintendents of navigation and entrusted with the job of sea transport. Between 1282 and 1391 ships laid a course closely following the coast, taking more than two months to reach Taiku (now Tientsin). Delays were caused by the sandy inshore shoals, by the strong on shore winter north wind and the southward flowing current (East China Cold Current). Later the transport fleet, consisting of 6-masted junks, sailed only between April and September each year, making two return trips. In so doing the adverse winter winds were avoided, and by sailing out almost due east from Liu-chia-kang, the southern terminus at the mouth of the Yangtze, and then turning north-west, the fleet was able to utilize to the full both the south-east monsoon and the northward flowing Kuro-siwo, thereby reducing the northward trip to a mere ten days.[1]

Although the Mongols were not natural sailors they nevertheless used sea transport to great effect in developing the spice trade between the Indies and Europe, via Persia and Egypt, during the short time they were in power. Ibn Battuta (1304–1378), the great Moslem traveller from Tangiers, reckoned Zaiton (Amoy) to be the greatest port of the world and was lost in wonder at the huge cargoes of the great four- and six-masted vessels, which sailed south carrying silk, porcelain, tea and camphor to the Malay Straits, where they picked up spices for their journey farther west to the Persian Gulf. When Marco Polo returned home, he used this route, landing at Hormuz, thence by land to Trebizond on the Black Sea. This trade undoubtedly had some influence in spurring on Portuguese sailors to their voyages of discovery later. The Venetians and Genoese, who were the principal European spice traders, carried spices from the Syrian and Egyptian ports to the rest of Europe. When Persia was converted to Islam and closed that door, Egypt remained the sole route of supply, and together with Venice and Genoa, foolishly used their monopoly to raise prices too high with a result that Portugal and Spain, and later England, France and Holland sought other routes.

Thus it was that eventually Portuguese sailors found their way to China and in 1557 were permitted to erect factories, i.e. small compounds containing offices, living quarters and warehouses on the isthmus of Macao. This was the beginning of a new era, which led to the invasion of the China seas by Western mercantile fleets, to the breakdown of Chinese isolation and with it the eventual crumbling of Confucian philosophy and the collapse of the last Imperial dynasty, the Ch'ing, in 1911.

[1] Chang Sun, 'The Sea Routes of the Yuan Dynasty, 1260–1342', *Acta Geographica Sinica*, Vol. 23, No. 1, 1957.

CHINA'S CAPITALS

Probably no other country in the world, except perhaps India, has moved its seat of government so often or so far as China. The movement has generally been logical and the siting has usually shown an attempt to resolve a tension between the rival claims of production and defence. There have naturally been many periods in Chinese history when the country has been divided and in consequence has had several capitals simultaneously.

In early post-neolithic times, when Chinese civilization was just developing, states and principalities of north China were small and isolated. Communications were poor and there was very little consciousness of, or contact with, neighbours. The focal centres or capitals of these small communities were based on local factors and had only local significance. Their armies were small and their resources for campaigning were not large enough to allow anything but very limited scope. As the population grew and states expanded they began to impinge on one another. It was not till late Shang times, when Chou began to encroach from the west, that the siting of a capital began to have really wide significance.

According to the *Shu Ching*, the Shang or Yin capital, when T'ang ascended the throne in 1766 BC, was at Po, but it was removed to Anyang in 1401 BC by P'an King. Tradition and classical writing attribute a jurisdiction to Shang rulers extending to the Yangtze, but it is probable that effective rule was very much more limited.

As we have seen, early Chinese civilization was fostered in the loess foothills and on the surrounding loess plain of Shansi. Anyang itself was on the loess plain in a bend of the river. Lying on the plain gave it the necessary protection against sudden attack and yet it was within reach of the wooded foothills for hunting. Hunting was then still an integral part of the people's livelihood, whilst later and for many centuries it remained the sport of the privileged classes.

It is worthwhile to pause here for a moment to note the significance of three place-names—Shensi, Shansi and Shantung. Shensi means 'west of the pass', that is, west of Tungkwan gorges, which lie below the confluence of the Wei and Hwang. Shansi means 'western hills' and Shantung, 'eastern hills', between which lies the flood plain of the Hwang Ho, where Chinese civilization developed.

The Chou, who were at that time barbarians occupying the Wei valley, issued through the Tungkwan gate and conquered Shang in 1122 BC. They destroyed the Shang capital at Anyang and removed the seat of government to Ch'ang-an, Shensi, on the right bank of the Wei Ho, whence they were better able to meet the threat of others rather similar to themselves farther west. Ch'ang-an stands athwart the easy entry from the west, which was later to become the Imperial Silk Route. However, the Chou dynasty ushered in the classical feudal era and it was not long before the rulers found it necessary to build a new capital which was more central for internal control. This they did at Loyang, which lies to the south of the Hwang-ho (Honan), above the flood plain. From here there is easy access to the Hwai basin and comparatively easy entry into the Han basin and so to the Yangtze, where the strong state of Ch'u was situated. The capital remained at Loyang until 220 BC.

The feudal period of Chou came to an end again as the result of western invasion. Chou had shrunk to a small state centred on Loyang, to which nominal allegiance was given even by the two most powerful feudal states. Ch'in and Ch'u. Ch'in, which had succeeded Chou in the Wei valley, first conquered Ch'u and then all the other smaller fry, and in 256 BC brought the whole country under one rule for the first time. Ch'in Shih Hwang-ti preferred to maintain his own former capital at Ch'ang-an. When his short but very momentous reign ended, his conquerors, the Han, reverted to Loyang as capital but very soon transferred their seat of government once more to Ch'ang-an, and there it remained from 220 BC to AD 25. This vacillation of capital between Ch'ang-an and Loyang reveals the tension between the rival strategic and economic claims of a capital. Whilst, as China grew in size and prosperity, the threat from the nomads of the north and north-west was increasing, at the same time the economic centre of the country was shifting steadily to the east and south. This tension increased with the rise of the Later Han (AD 25–220), when Chinese rule extended into Sinkiang and beyond the Pamirs. Trade along the Silk Route was most active. One would therefore expect, that Ch'ang-an would continue as the imperial seat, but expansion southward was of even greater significance, and once again Loyang became the governmental centre.

With the fall of the Later Han in AD 220 there follows a long period of disunity (AD 220–581). The empire was first divided into three kingdoms. The northern and central provinces, under the King of Wei, had Loyang as capital. The King of Wu governed the southern provinces of the Yangtze valley from Nanking, and the King of Shu, the western provinces (Szechwan) from Chengtu. At one time there were as many as sixteen states, each with its own capital city. From 317 to 589 a succession of small independent kingdoms (E. Chin, Sung, Ch'i and Liang) had Nanking (then known as Chien Kan) as capital. It was not until the Sui (589) and T'ang dynasties (618) that the country was once again united. T'ang came down from

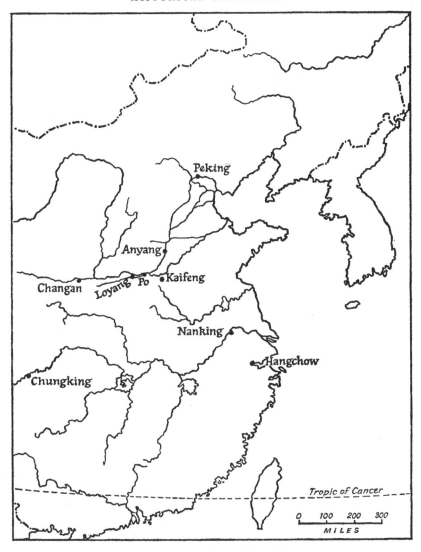

FIG. 32. China's imperial capitals

Shansi to conquer the land. Their capital had been at Taiyuan, but they followed the lead of Sui and moved to Ch'ang-an. They also maintained an eastern court at Loyang and built the New Pien Canal to link the two cities with the Yangtze valley.

China enjoyed united rule for 680 years between the end of Chou (220 BC) and the fall of T'ang (907). During this time Ch'ang-an was capital

93

of the Empire for 530 years, thus emphasizing the importance of the western gate during those earlier years. It was not until the economic heart of the country shifted into the Hwai and Yangtze valleys that Ch'ang-an lost its ascendency. It is instructive to note that at no time was the capital sited farther west than Ch'ang-an, even when in Han and T'ang times Chinese rule extended into Sinkiang and Turkestan. The 15 in. isohyet lies not far west of Ch'ang-an, and one quickly passes from true Chinese agricultural economy into nomadic pastoral way of life as one goes westward.

We have seen that it was not until the T'ang dynasty that the Yangtze valley and Nanling (South China) were fully incorporated into China, and the Yangtze valley became the key economic area. Towards the end of the T'ang era, threat of nomadic invasion shifted from the north-west to the north-east and in 907 the dynasty fell to the Kitan attacks from Manchuria. Then followed a further long period of disunity (907–1227). The Kitan kingdom of Liao was established in the north, having Yenching (Peking) for a capital, while the true Chinese kingdom of Northern Sung moved to Kaifeng to hold the gate to the south. Later, under Mongol threat from the north, Southern Sung emperors fled south and made Hangchow their centre, Nanking being too vulnerable. Until AD 591 Hangchow had been only a small fishing village, but after that date it developed rapidly as a trade centre and eventually became a great city of 2 million people with a wall which was 24 miles in length. Hangchow suffered temporarily when the Mongols under Kublai Khan defeated Sung but it quickly rose again although never as a capital. It was Hangchow which so excited the admiration of Marco Polo. In 1852 it suffered very severely at the hands of the Taiping.

In 1280 China was once again a united country under the alien rule of the Mongols (Yuan Dynasty). Jenghis Khan defeated the Kitans and captured Yenching in 1234. It was raised to the status of imperial capital in 1271 when Kublai Khan rebuilt it, naming it Cambaluc. When the Ming drove out the Mongols in 1368 they favoured Nanking as their centre for thirty-four years, but in 1402 they moved to Peking. The capital remained there throughout the rest of the Ming and the whole of the Ch'ing (Manchu) dynasties and to the present day, except for the twenty-two years of Kuomingtang rule between 1927 and 1949, when Nanking and Chungking shared the honours.

Peking served well the needs of the three dynasties Yuan, Ming and Ch'ing. Standing near the Nankow Pass, it proved a good centre from which Kublai Khan could govern China itself and from which, at the same time, he could keep in touch with his vast steppeland territories to

PLATE III: Although modern methods of irrigation by huge dams and mechanical pumps are being developed, traditional methods are still very generally in use. Here water is being raised from wells by human and by donkey power for irrigation on the North China Plain.

PLATE IV: Raising water by 'dragon skeleton water wheels', flooding the rice fields already planted out in Shantung.

PLATE V: Water buffalo are used for ploughing, particularly in the south. Note the two big 'kangs' in the foreground, used for maturing night soil—the main fertiliser in all China.

the west. For the Mings, although they drove out the Mongols from China but never subdued them in their own lands to the north, Peking served to hold the gate against them. The Manchus entered and conquered China mainly through the narrow coastal plain of Jehol at Shankaikwan. Again Peking was admirably situated for the simultaneous government by the Ch'ing emperors both of China and their own home territories in Manchuria.

China's Capitals

Ch'ang-an (Sian)

1122–255 BC	Chou Dynasty. Known as Haoking	
221–206 BC	Ch'in Shih Hwang Ti	Imperial
206 BC–AD 25	Former Han	Imperial
AD 316–329	Former Tsiao	
352–383	Former Tsin	
384–417	Later Chin	
534–544	Western Wei	
577–581	Northern Chou	
581–619	Sui	Imperial
619–907	T'ang	Imperial

Loyang (Honan)

770–249 BC	Eastern Chou	
AD 25–196	Later Han	Imperial
220–265	Wei	
265–311	Western Chin	
431–534	Later Wei	
924–938	Later T'ang	

Kaifeng (Pienliang; Pienking)
With the fall of T'ang there followed five dynasties of five military despots:

907–924	Later Liang	
924–938	Later Tang	
938–947	Later Chin	
947–951	Later Han	
960–1127	Northern Sung	Imperial

Hangchow (Linan)

1127–1280	Southern Sung	Imperial

Nanking (Chienyi; Yintien)

229–280	Wu	
316–420	East Ching	
420–479	Lu Sung	
479–520	Tsi	
520–557	Liang	
557–589	Chen	
942–965	Southern Tang	
1368–1402	Ming	Imperial
1850–1861	Taiping	
1927–1949	Kuomingtang	

Peking (Yenching; Cambaluc)

937–1123	Liao	
1150–1234	Chin	
1271–1369	Yuan	Imperial
1402–1637	Ming	Imperial
1646–1911	Ching	Imperial
1911–1927	Republic	Imperial
1949–	People's Republic	Imperial

THE IMPACT OF THE WEST

Before embarking on the economic and social development of China it is necessary, as a background to that development, to follow the course of the geo-political revolution which has rent China during the last hundred years and to examine some of its ingredients.

To all intents and purposes China has rested in self-sufficient isolation over the past three thousand years. There have been invasions from the north, but the conquerors have quickly been assimilated. It is also true that Buddhism infiltrated, mainly over the land routes via Khyber and Karakoram through Sinkiang to China, but Buddhism was also absorbed, adapted and transmuted to become an integral part of Chinese civilization. Although Chinese rulers have seldom favoured Buddhism at the expense of Confucianism, this integration is the main reason why Buddhism has suffered less interference and persecution at the hands of the present Communist government than the two religions of Islam and Christianity, which have remained alien.

The Christian religion made its first entry in its Nestorian dress, carried by Syrian monks in the seventh century, but it made no deep or lasting impression. There was some contact with Rome during the Yuan dynasty through such travellers as William Rubruck, Friar John and Marco Polo. Kublai Khan ordered a great conference of religious leaders, Hindu, Buddhist, Christian and Moslem, in an endeavour to find a faith for his people. He sent a request to the Pope in 1269, asking for a deputation of one hundred learned men who would expound the Christian faith. This request was never answered, which is the more surprising since Christendom at that time had hopes of Mongol assistance against Moslem aggression on Europe's eastern borders. It is interesting, although perhaps not very profitable, to speculate on the course of history had these men been sent.

The first real impact of Christianity on China was made by the Jesuits in the seventeenth and eighteenth centuries. Led by such men as Xavier (1549) and Ricci (1601), the Jesuits were a body of learned and dedicated men, who set out to convert the intelligentsia, the fountain-head rather than the masses, to Christianity. They achieved a great measure of understanding of Chinese culture and scholarship and attempted an integration of Confucianism and Christianity. Their learning, especially their science, earned them respect and favour in court and scholarly circles, and for a long time they enjoyed a monopoly of the missionary field. It is doubtful, however, how deep an impression they really made. The Manchus were ready enough to accept their scientific contribution, which was very considerable. Much more doubtful was their acceptance of religion. In fact, Emperor K'ang Hsi repudiated it in no uncertain terms: 'As to the Western doctrine, which exalts the Lord of Heaven, it is opposed to our traditional teaching. It is solely because its apostles have a thorough knowledge of mathematical sciences that they are employed by the State. Be careful to keep this in mind.' Later the appearance of Dominican missionaries, who came with the idea of the conversion of the masses and an overthrow of a 'heathen' culture, diametrically opposed to the Jesuit idea of integration, led to bitter strife between them. K'ang Hsi sarcastically remarked: 'You Christians go to a lot of trouble, coming from afar to preach opinions about which you seem anxious to slit each others throats.' The Pope in 1742 joined in condemnation of the Jesuits' attempt at integration of Confucianism and Christianity. It was not until 200 years later that a papal edict rescinded this and stated that Confucian rites and Roman Catholic doctrine were not mutually exclusive.[1] Manchu patience was

[1] For a full discussion of the spiritual and intellectual impact of the West see A. de Riencourt, *The Soul of China* (London 1959), Chapters 8, 9, 10 & 11

exhausted by this strife in the eighteenth century and Christian missionary work was prohibited.

For nearly a century and a half China was ruled by two of its ablest emperors, K'ang Hsi (1661–1722) and Ch'ien Lung (1736–1796) and enjoyed one of its most glorious periods. The bounds of its empire were wider than at any previous time except that of the Yuan dynasty. Its culture, art, literature and crafts were both admired and envied by the West. Nevertheless, that civilization was static and petrified and was already showing signs of crumbling. At the turn of the eighteenth and the beginning of the nineteenth centuries Western traders, mainly British, were increasingly pressing for more trade with China. The official Chinese attitude was to discourage external trade on the grounds that China herself had all things needful within her borders. Trade was therefore confined within very narrow limits. Canton alone was the port of entry. There, a few 'factories', i.e. combined warehouses, offices and living quarters were permitted to foreign merchants under strict control and as a result smuggling became prevalent. Western truculence and Chinese arrogance clashed. Chinese prohibition of the importation of opium, the most lucrative article of trade, led to wars which resulted in Chinese defeat in 1840.[1] This date marks the beginning of the western 'break in' and the beginning of a century in which western individualism, nationalism and capitalism imposed themselves on the autocracy and universalism of Confucian China. For one hundred years the Western powers with their vast superiority in military and economic strength forced open Chinese ports, claimed extra-territorial rights, schemed for spheres of influence and concessions in which to invest their capital, and generally rode rough-shod over a decadent and impotent China to its mortification and humiliation.

With the opening of Chinese ports, Christian missions, both Catholic and Protestant, again moved in. In contrast to the earlier Jesuit attempt at Christianization, their efforts were directed almost entirely to the common people, the *peh hsin* and at first they made very little impact on the intelligentsia. While their early work was in the field of evangelism, they moved increasingly into first medicine and then education. Their influence in these fields has been far-reaching and revolutionary. The work of mission hospitals, doctors and nurses gave birth to the present vast and ever-growing medical services, which, while bestowing many benefits, are not a little responsible for the present population problems. Early evangelists, intent on the spread of the Gospel to the masses, translated the Bible into *kwan hwa*, i.e. writing which followed closely the spoken word in contrast

---

[1] See the section on Foreign Trade on p. 200.

to the abstruse literary *wen hwa*, intelligible only to the literati. Mission schools introduced Western learning, Western democratic ideas, Western history and Western science. Large numbers of students have gone abroad to Europe and America for further education. The inevitable result has been a surge of revolutionary ideas in many fields.

In spite of the enormous amount of altruistic and self-sacrificing work done by Western missionaries and in spite of the vast amount of capital expended in this work, the final result has been a discrediting of Christianity in the eyes of the Chinese people. Christianity has come to be identified in the Chinese mind with Western culture and so with Western imperialism. So often the Christian missionary was made use of in the political game, although seldom so consciously and so cynically as did Napoleon, who is reported to have said: 'Religious missions may be very useful to me in Asia, Africa and America, as I shall make them reconnoitre all the lands they visit. The sanctity of their dress will not only protect them but will serve to conceal their political and commercial investigations.'[1] Throughout the nineteenth century the expanding field of trade went hand in hand with the expanding mission field, sometimes preceding it but often following in its wake and making use of some incident, maybe the murder of a missionary or an arrogant action by the Chinese, as a pretext to exact further concessions.

Physical intrusion in the nineteenth century by the west, with which we must also associate Japanese aggression against China, was marked by two wars of 1840–42 and 1856–60 between Great Britain and China, which opened the land to trade and to missions; by a civil war, the T'ai P'ing Rebellion, which had its roots in peasant discontent combined with obscure fanatical Christian beliefs; by a war in 1894 between Japan and China; and finally by the Boxer Uprising, fostered by the reactionary Dowager Empress, Tz'u Hsi ('Old Buddha') in a bloody and foolish attempt to oust the foreigner and all his ways. All these opened the way for Westerners (British, French, Germans, Russians, Americans, Scandinavians) and Japanese to trade, to develop by investing capital and to exploit.[2]

Significant and far-reaching as were these events, even deeper was the spiritual and intellectual impact made by this entry. During this same period, Japan had had to face this same virile, self-confident individualism from the West with its new techniques and democratic ideas, but she was strong enough to meet the challenge. In judo fashion she accepted and

---

[1] H. G. Wells, *An Outline of History* (London 1920), p. 793.

[2] For a full history of these events see K. S. Latourette, *The Chinese: Their History and Culture* (London 1956).

adapted the new forces to suit her old religion and philosophy without significantly changing their essence. China, on the other hand, was too weak and decadent to do more than acquiesce or to resist passively. The philosophy of individualism, *laissez-faire*, nationalism and natural science of Hume, Bentham and Descartes entered through all channels. The old Confucianism was for a while discarded in favour of this new philosophy, which commended itself by the very power, wealth and vitality of its exponents, but not for long. It could not satisfy the innate Chinese desire for universalism and collectivism, which has supported their civilization for three thousand years. Moreover, Russia's defeat at the hands of Japan in 1904 and the spectacle of the First World War revealed a moral bankruptcy which could not but discredit the West. Add to this the intense hatred that was felt by most of the intelligensia for that Western imperialism, which had so humiliated their country, and it is not surprising that Chinese thinkers of the early twentieth century turned to German philosophers for satisfaction. This they found in the writings of Hegel, which appealed at once to their universalism and to their sense of history. It further enabled them to retain Western science and industrialism while repudiating the rest. It needed only Lenin's adaptation of Marxism to the needs of the Orient under the banner of 'Marxism of the Era of Imperialism' for Marxist-Lenin dialectical materialism to be embraced by the revolutionaries and for it to become the creed under which the Communist Party eventually secured control of the country.

CHAPTER THREE

Economic and Social Geography

POPULATION AND ITS PROBLEMS

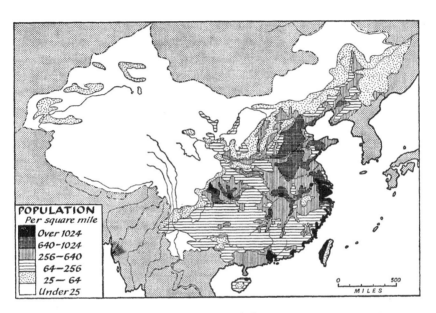

FIG. 33. Population

Statistics

Reliable statistics are essential in any planned society. Without them those in authority must work by rule of thumb and no intelligent planning can be undertaken. First and foremost of the planners' statistical needs is a reasonably accurate recording of the number of human beings to be served.

Until the census of 1953, China has had to be content with estimates of population which, it is universally agreed, are unreliable to a degree. In the past these estimates have been based variously on the number of households, the average food or salt consumption, the amount of postal material, etc., and not unexpectedly, the results have shown great disparity, as the figures overleaf reveal:

Source	Year	Population
Imperial Government	1910	330,000,000
Walter F. Willcox	1926	323,000,000
Post Office	1926	485,508,838
Maritime Customs	1931	438,933,373
Ministry of the Interior	1931	437,380,000
Chou En-lai	1950	487,000,000

In 1953 the government of the People's Republic undertook the first real census. In view of the vastness of the country and the number of trained enumerators who would have been required, a single day count was not attempted. A *de jure*, i.e. habitually resident, basis was favoured rather than the *de facto*, i.e. actually present, as used in Great Britain, since it was considered that this would give a more accurate result and also would give a more useful picture of the economic, social and cultural make-up.[1] The task was stupendous and it is very difficult to estimate the degree of accuracy that was attained, but there is agreement amongst most informed opinion that the 1953 figures provide a working basis with a margin of error of perhaps 25 million or 4 per cent. The margin of error in other statistics, such as agricultural and industrial production, is probably much greater. A second census had been proposed for 1963, but nothing further has been heard of it, nor apparently are any preparations being made to carry it out. Unfortunately there has been a serious setback in statistical data since 1958, when the State Statistical Bureau, which was beginning to make good progress, came under the criticism of over-enthusiastic Communist Party cadres, from which it has not yet fully recovered.

Composition

Nearly 94 per cent or some 547 million of the population is Han, which term is used here to include both the northern Han Jen and the southern China T'ang Jen or Cantonese. They are found almost entirely in the area known as China Proper and the North-East (Manchuria), but are distributed very unevenly within this area, being concentrated densely in the river basins and rich alluvial plains, notably the North China plain, the Red Basin, the Lower Yangtze Basin and the Lower Si Kiang Basin. Some rural areas have a density of more than 2,000 people to the square mile. This is the great homogeneous mass of the Chinese people, having the same

[1] S. Chandrasekhar, *China's Population* (Hong Kong 1959).

ideas, the same traditions and culture and, although there are many dialects so different from one another as to make one region's speech incomprehensible to another, the same written language. Apart from the Hui, the Moslems of the Kansu area, there are no religious divisions or antagonisms, such as are met with in the Indian sub-continent, and this Moslem group numbers only 3½ million.

We have seen that the Chinese (Han Jen) in their southward expansion from the Lower Hwang Ho into the Yangtze Basin and eventually into the Si (West) River Basin in the early centuries AD, met with, and often absorbed, the aboriginal tribes which they encountered. However, many resisted the invaders, falling back before them and taking refuge in the more inaccessible highland parts, rather as the Britons fell back to the Welsh highlands and the more remote south-west peninsula of England before the Anglo-Saxons. These aboriginal tribes of China have continued to maintain their identity and individuality through the centuries down to the present day. Thirty-three such peoples are listed by name as National Minorities in the census of 1953 and altogether number 35,320,360 or some 6 per cent of the whole population. Those which number more than one million are listed below:

China's National Minorities

Minority	Region	
Chuang	West Kwangsi Province	6,611,455
Uighur	Sinkiang Uighur Autonomous Region	3,640,125
Hui (Moslems)	Kansu and Chinghai	3,559,350
Yi	Szechwan-Yunnan borders	3,254,269
Tibetan	Tibet, Chamdo and Chinghai	2,775,622
Miao	Kweichow and West Hunan	2,511,339
Manchu	Widely distributed	2,418,931
Mongolian	Inner Mongolian Autonomous Region	1,462,956
Puyi	South-west Kweichow	1,247,883
Korean	Yenpien Korean Autonomous *chou* in Kirin	1,120,405

Population Distribution by Provinces (1953 Census)

China Proper	Population
Szechwan	62,303,999
Shantung	48,876,548
Honan	44,214,594
Kiangsu	41,252,192

Hopei	35,984,644	
Kwangtung	34,770,059	
Hunan	33,226,954	
Anhwei	30,343,637	
Hupeh	27,789,693	
Chekiang	22,865,747	
Kwangsi	19,560,822	
Yunnan	17,472,737	
Kiangsi	16,772,865	
Shensi	15,881,281	
Kweichow	15,037,310	
Shansi	14,314,485	
Fukien	13,142,721	
Kansu	12,928,102	
		506,738,390
North-East		
Liaoning	18,545,147	
Heilungkiang	11,897,309	
Kirin	11,290,073	
		41,732,529
Inner Mongolian Autonomous Region	6,100,104	
Jehol	5,160,822	
Sinkiang	4,873,608	
Sikang	3,381,064	
Chinghai	1,676,534	
Tibet and Chamdo	1,273,969	
		22,466,101
Three Municipalities		
Shanghai	6,204,417	
Peking	2,768,149	
Tientsin	2,693,831	
		11,666,397
Total for Mainland, including National Minorities		582,603,417

Of this total 505,346,135 or 86·74 per cent are stated to be rural, that is, living in hamlets, villages or small towns of not more than 2,000 population or in larger towns in which more than half the people rely on agriculture for a livelihood. The remaining 77,257,282 or 13·6 per cent are classed as urban.

Age Composition

China today is a young people's country. Census figures show that 86·5 per cent of the population in 1953 was below 50 years of age and that 41·1 per cent was 17 years or less. USA figures for 1960 were: 0–14 years (20·8 per cent); 15–59 years (63·9 per cent); over 60 years (15·3 per cent); gross reproduction rate 1·75 per cent.[1]

This youthfulness of the nation cannot but have great influence on the economic development of the country as it pursues its policy of industrialization in the next few decades, since it means that a very large proportion of the population will be of productive age.

Another interesting demographic fact is that, contrary to present Western experience, males outnumber females, there being 107·7 men for every 100 women. S. Chandrasekhar gives the following table[2]

Age Groups	Sex Ratio: no. of Males per 100 Females
0	104·9
1—2	106·2
3—6	110·0
7—13	115·8
14—17	113·7
18—35	111·5
36—55	106·8
55 and over	86·7

It will be seen that the disparity between the two sexes is becoming less. This reflects a change in woman's status: the female baby today is less at a disadvantage as compared with the male than in former years, when in times of stress, female infanticide was practised.

Modern medicine and public hygiene are responsible for great changes in death rates in China in the last quarter of a century and particularly in the last decade, during which the People's Government has concentrated on the spread of knowledge of public hygiene. S. D. Gamble, investigating Ting Hsien between 1933 and 1936, quotes a birth rate of 40 per 1,000 and a death-rate of 27 per 1,000, giving a natural increase of 13 per 1,000 or

[1] *U.N.O. Population Studies*, No. 26, 1956. 'The Ageing of Populations and its Economic and Social Implications.'

[2] S. Chandrasekhar, op. cit., p. 43.

1·3 per cent.[1] It is claimed that the death rate fell from 17 to 12 per 1,000 between 1953–6 and that the infant mortality dropped from 200 to 81 per 1,000 between 1949–56. The birth-rate remains fairly constant at about the rate of 3·7 or 3·8 per cent. As a consequence it is estimated that there is an annual natural increase of 2·4 per cent or more. If this is true and if the trend is maintained, it will mean that by 1975 China will have a population of some 800 million and by 1990 well over 1,000 million or about one third of the world's people. Expectation of life in China, which was estimated as low as 27 at the turn of this century was stated, in 1958, to be 54 years as compared with Japan's 66 years, Taiwan's 55 years and India's 35 years.

Overseas Chinese

The Chinese Government is insistent that Taiwan is an integral part of the People's Republic of China and claims the island's population as its own. Figures published by the Nationalist (Kuomintang) Government in 1951 showed a population of 7,591,298.

Contrary to normal international usage, Chinese who have settled abroad and students studying abroad at the time of the census, have been included in China's census. According to the Commission of Overseas Chinese Affairs these number 11,743,320. Professor S. G. Davis lists the distribution of the bulk of these as follows:

Country	
Thailand	3,500,000
Malaya	3,013,000
Indonesia	1,598,000
Indo-China	1,221,000
Singapore	861,000
Burma	360,000
Sarawak	164,000
Philippines	154,000
North Borneo	83,000
Japan	38,000
Hawaii	30,000
Korea	18,000
India	14,000
	11,054,000

[1] S. D. Gamble, *Ting Hsien. A North China Rural Community* (Institute of Pacific Relations, 1954).

Population Problems

The present Communist government in China has followed a rather vacillating policy with regard to population. When it first came into power in 1949 it was quite emphatic in its denial of any truth in the Malthusian doctrine. It was optimistic that it could meet any increase in population by a more than proportional increase in production, and in fact welcomed the increase in numbers. Towards the end of 1954 some doubt set in when production did not increase as fast as was hoped. Some measure of birth-control was therefore mooted. Shao Li-tze, in a speech to the China National People's Congress, stated 'It is a good thing to have a large population but, in an environment beset with difficulties, it appears that a limit should be set'. For a short while whole-hearted official support was given to the birth-control campaign, but within a year, with the growth of production, official blessing was again transferred to increasing numbers.

If she is not willing to curb these growing numbers, China can meet their needs in two ways, namely by migration within her own borders or by increased production proportional to, or greater than, the increasing population.

Migration holds some possibilities. There are two areas to which major movements could be made, namely, the North-East and the north-west. The high Tibetan plateau is left out of consideration as being, in the present state of knowledge, beyond man's power to colonize or utilize to any large extent. Although the North-East has been settled extensively in this century after Manchu prohibition of entry had been lifted, there are still vast areas, particularly in Kirin and Heilungkiang, awaiting agricultural development and still rich mineral deposits so far unexploited. This region can absorb many millions profitably.

The unsettled, unutilized land of the North-west is of far greater extent than the North-East but it presents very serious problems of development. It consists mainly of a number of inland drainage basins, the centres of which are usually true desert or very arid land. The peripheral areas are more fertile, usually semi-desert or steppe, the home throughout the centuries of thinly dispersed nomadic tribes. Some of this could be turned into arable land if water were brought to it in sufficient quantities. This is already being done on a large scale in Sinkiang and Chinghai, but it requires the application of much capital and moreover incurs the danger common to all dry lands coming under irrigation—the danger of alkaline accumulation.

There is a third area which invites Chinese occupation but which is outside Chinese political boundaries: the vast lowlands of the Indo-China peninsula. These lands are comparatively under-developed and sparsely populated. Apart from fairly dense concentrations around Bangkok, Saigon and Hanoi, most rural areas have well under 100 persons per square mile. The climate of the region is very similar to that of south China and there is virtually no relief barrier. The figures of overseas Chinese given above show that large numbers have settled here in the past. It may well be that Chinese pressure in this region today stems from socio-demographic as much as from political and ideological causes.

The other way of meeting the needs of China's growing population is to increase the yield from the land already occupied. In the long run, this is the method on which all countries and peoples must rely, since un-occupied lands are limited. Arable land in China Proper has been very in-tensively cultivated for centuries. Can cultivation be further intensified to give higher yields, and is it possible to bring areas now uncultivated under the plough? An increased yield of well over 2 per cent per annum is called for merely to maintain present living standards, if the estimated natural increase of population is reasonably correct. There are, in fact, many ways in which food production can be increased, and all of them are being attempted. They are listed and commented on briefly below. Each will be given fuller treatment in its proper place in later pages.

1. FLOOD CONTROL

The amount of land actually yielding a harvest in any one year can be increased by flood control. Scarcely a year passes but one or all of the great rivers of China, the Hwang Ho, Yangtze, Hwai and Si, break their banks at some point and large areas are inundated and the crops destroyed. In years when there is abnormally heavy summer rainfall, these inundations may be catastrophic as they were in 1931 and 1954 in the Yangtze and Hwai Basins. It is estimated that 34,000 square miles of the middle and lower Yangtze Basin were flooded in 1931. Most of this was arable land, and at that time under crops, mainly rice. We may reckon that some 25,000 square miles or 16 million acres or approximately 100 million *mow* were lost to cultivation in that year. In terms of food this meant a loss of 250 million *tan* (1 *tan* equals 133 lb.) since one *mow* produces an average of $2\frac{1}{2}$ *tan*. Losses of this calibre have been the rule rather than the exception throughout the centuries. Grandiose schemes and strenuous efforts are now being made to prevent these catastrophies and thus increase the food supply. Afforestation and dam construction, which play so large a part in flood control, are dealt with later.

2. IRRIGATION

A considerable amount of new land has been brought into cultivation by the extension of irrigation, notably in the drier regions, such as Sinkiang and Kansu. Even by 1955 it was claimed that 33,000 hectares (approximately 80,000 acres) of new land had been brought under the plough in Sinkiang alone. Many other similar schemes have been put in hand but probably the thousands of small, local efforts throughout the country are of greater importance. By works damming up small heads of water, a perennial supply, instead of a seasonal supply, has been brought to already cultivated lands, thus enabling them to double crop the fields and virtually to double the acreage.

3. GRAVES

The social revolution effected by Communism has enabled a reform, long advocated by agriculturalists, to be carried through. Confucian filial piety has required great reverence for ancestors. This has led, throughout the country, to leading members of the family being buried in favoured spots, often in the middle of the best land as a mark of special reverence and piety. Lossing Buck has estimated that 1·9 per cent of Chinese farmland was devoted to this purpose.[1] It is reported that nearly all such graves have now been removed to less productive sites or reburied more deeply so that they can be ploughed over. If this is so, it will have provided land nearly sufficient to meet one year's increase in population.

4. IMPROVED FARMING TECHNIQUES

Chinese farming methods have changed very little over the centuries. The peasant has tended his plot with assiduous care and patience and has produced from the land all that could be produced by those methods. It is claimed that production can be trebled or quadrupled by the application of modern knowledge and techniques—composting, artificial fertilizers, deep ploughing, seed selection, pest control, better tools, etc.

5. COMMUNICATIONS

By the improvement and extension of communications and transport a better distribution of the foodstuffs produced can be effected. Much of what has been produced in the past has been wasted, or at least not put to its optimum use, particularly in years of scarcity and famine.

[1] Lossing Buck, *Land Utilization in China* (Shanghai 1937).

6. FISHERIES

Considerable food supply is obtained throughout south and central China from fish ponds, which are to be found in nearly every village and hamlet. These are stocked each year and fed from the refuse of the village. The yield from these ponds could be greatly increased by the application of careful and scientific methods. Where this is done at present the return per acre is double that derived from paddy fields.

7. INDUSTRIAL DEVELOPMENT

In the past most farms in China have had far more labour than could be economically used during the greater part of the year. Industrial development would serve to draw these redundant land workers into the towns. This would result in an increase in the production of goods and a rise in the standard of living. It is difficult to see, however, how this in itself would result in an increase in food production, except in so far as industry would contribute certain materials, such as chemical fertilizers, which would add to the fertility of the soil.

These are the main ways, in the present state of knowledge, available to the Chinese to meet the needs of their growing numbers. Whether these methods are adequate or not is a matter of opinion. Taking South-east Asia as a whole, it would appear that they are insufficient since it is estimated that there is less food per head available in that part of the world than there was ten years ago. There are exciting possibilities, some of them in an advanced experimental stage—synthetic foods, the use of green algae *shlorella*, yeasts, seaweeds, the cultivation of the sea bed—which might revolutionize food production and possibly lay the Malthus bogey for some generations.

ECONOMIC DEVELOPMENT BEFORE 1949

In treating the subject of the economic development of China we have found it convenient to sub-divide it into two very unequal parts in time: ante 1949 and post 1949. The period before 1949, stretching right back into the Shang and Chou dynasties, gives us, as it were, the physical and human material from which the present revolutionary structure is being carved, and for this reason it is essential that we give it careful scrutiny. In looking at both periods we need especially to bear in mind Owen Lattimore's warning that, while the factors of physical geography, climatic stimulus and the character of environment must be given full consideration in assessing any given situation, mature understanding requires also that proper weight be given to the dynamics of social groups. In view of Communist repudiation of determinism and its claim to mastery of the environment, any treatment of the economic development of the country in the period after 1949, the year in which the People's Government was established, must pay particular attention to human activity as well as to natural conditions. Far-reaching and fundamental changes in economic geography have taken place since 'liberation'. Nevertheless, much of the experience and ingrained habit, formed over 3,000 to 4,000 years remains and has conditioned to a considerable degree the changes made since 1949.

Agricultural Development

From the break-up of feudal society (221 BC) until the intrusion of modern Western culture and ideas, China has had a social pattern which had accepted values and social gradations stemming essentially from Confucian philosophy. The recognized social scale placed the scholar-administrator at the top and the farmer, craftsman, merchant, soldier and criminal in descending order below him. That is not to say that the soldier was not often in the place of power, but inevitably, inexorably, the scholar-administrator each time superseded him. Equally inevitably, the scholar-administrators became the landowner and landlord class with a vast peasantry below them, supplying their needs. It was the intention, the very basis, of Confucian philosophy that the scholar should be the 'princely man': just, upright, benevolent and revered for his learning; and, perhaps more often than is generally credited, the scholar approximated to this ideal. Certain it is that, through the centuries, scholarship, which incidentally led to office, was held in high esteem. The scholar-landlords were numerous, and for the most part did not hold enormous estates.

Generally, as each successive dynasty became decadent, there was a concomitant deterioration in relations between landlord and tenant. Landlords became more grasping and oppressive, and taxes and rents pressed more heavily on the peasantry. Nearly every change of dynasty has been sparked off by peasants revolting against intolerable conditions, clamouring for reform. The first half of this century has been one long and continuous tale of political unrest, during which no less than three considerable revolutions occurred, of lawlessness and banditry, of the rapacity and ruthlessness of war lords, of ever increasing taxation. It is in this setting that agricultural production and landlord-tenant relations up to 1949 must be seen.

According to an estimate made by the National Agricultural Research Bureau in 1917, 46 per cent of the farmland of China was held by owner-farmers; 24 per cent by farmers who owned part of the land they worked and rented part; 30 per cent by tenant farmers. The percentage of tenant-farmers increased very considerably during the succeeding decades. The proportions of land holding differed quite markedly between north and south. It is estimated that north of the Hwang-ho more than 80 per cent were owner-farmers. There was increasing part ownership and tenancy farther south. Between 20 and 40 per cent fell in these categories in the Yangtze valley below the Gorges, and more than 40 per cent south of Nan Ling, i.e. Kwangtung and Kwangsi. Szechwan contained the highest proportion, there being over 50 per cent tenant-farmers in that province. One influence of Western culture and contacts was to increase the flow of landlords into towns and so increase absentee landlordism with all its attendant shortcomings.

Security of tenure of the land was fairly well assured so long as rents were paid. Rents varied both in basis and amount. It was usual to fix on one of two bases, share rent or crop rent. With the former, the landlord took usually half the crop and so shared to some extent in both years of prosperity and adversity. Years of adversity naturally pressed more heavily on the tenant who was living near subsistence level than on the landlord, who had some reserves. Crop rent was a fixed amount, in weight of a stipulated crop, usually rice in the south and wheat in the north. This had to be paid regardless of the harvest. On average this worked out at between 45 and 50 per cent of the farmer's harvest. Both bases pressed heavily on the peasant farmer. Some peasant holdings were let on a cash basis, the annual rent being about one tenth the cash value of the land. This class of tenure formed only a small proportion of the whole. It was increasing and was particularly favoured by the absentee landlord.

The average size of farm for the whole country was about 24 *mow* or 6 acres. It was estimated that 36 per cent of the farms were less than 10 *mow*; 26 per cent between 10 and 29 *mow*; 25 per cent between 30 and 49 *mow*; 10 per cent between 50 and 90 *mow* and only 6 per cent more than 100 *mow* or about 17 acres. There was a considerable variation, as might be expected, between north and south. The largest farms were to be found in the North-east (Manchuria), where the average was about 120 *mow* (20 acres). On the dry-farming, wheat-growing North China Plain the average was 45 *mow* or 7·5 acres, while in the rice growing lands of the south it was much lower, being only 12–18 *mow*. The farm land per head of population amounted to only 2–3 *mow*. The ever growing population during the last century led to greater and greater land scarcity and therefore to increasing competition for it. Rents rose constantly, to meet which the land was ever in greater danger of being overworked and impoverished. The number of small tenants who were forced by this to give up their holdings and to live as agricultural labourers grew steadily.

Although holdings were so small, seldom could one find a compact farm of any size. Most farms consisted of scattered strips and patches, often half a mile or more apart, between which the farmer had to divide his attention and to hump his plough.

Even greater handicaps than those of land scarcity and scattered fields were the lack of credit and the poverty of markets. In company with farmers the world over, the Chinese peasant faced the problem of slow turnover. Living as he did with little or no reserve and so often on the verge of subsistence, the long wait between the preparation of his fields and the sale of the crop was a very serious matter for him. He was always in need of capital. Generally he was so pressed that he had to sell his cash crops (usually wheat or rice) immediately on harvesting. This meant that he was selling cheaply on a buyers' market. Often he found that he had to buy back rice or wheat for the needs of his family in the spring when prices had risen sharply. This, together with the socially imperative expenses connected with celebrations at birth, marriage and death, led the peasant into the hands of the moneylender, who often was also his landlord. A very moderate rate of interest was 2 per cent per month; 100 per cent per annum was not unusual.[1]

Owing to ignorance and to poor communications the peasant's market was a very restricted one, seldom reaching beyond the nearest large town and more usually confined to his own village, where he would be largely at the mercy of visiting merchants. Both communications and means of transport were so poor that, except in a few favoured areas, there could be no wide market. Wheeled cart and pack animal were used over the dust and mud tracks of the north, and wheelbarrow and coolie pole were used over the paths of the south. Very little mechanized transport was in use before 1949. In consequence of this isolation prices varied enormously between places no great distance apart. Up till 1949 very little use was made of co-operative societies, which might have eased some of the difficulties of credit and marketing.

[1] For a comprehensive and objective discussion of the position and problems of agriculture and industry in the earlier part of the century, see R. H. Tawney, *Land and Labour in China* (London 1932). For intimate and detailed analysis see Hsiao Fei-tung, *Peasant Life in China* (London 1947).

In spite of all handicaps of land scarcity, insecure or oppressive tenure, debt and poor markets, the Chinese peasant has deservedly gained a world-wide reputation for his attachment to, and his love of, his land, and for the meticulous and indefatigable care he lavished upon it. Given his knowledge and the tools and material available to him, it is doubtful whether much more per acre could have been wrested from the soil. He was, however, severely handicapped by ignorance for which all his care and attachment could not compensate. His techniques were those dictated by countless generations before him. He planted and reaped according to his agricultural calendar. He knew little or nothing of seed and crop selection or how to counter the many pests which yearly ruined his harvest. His tools were primitive, consisting almost exclusively of heavy hoes, clumsy, heavy wooden or bamboo rakes and harrows, crude and cumbersome ploughs, which often did little more than scratch the surface, and ingenious but laborious caterpillar waterwheels. While he was constant and faithful in feeding the soil, using carefully matured night soil and whatever animal manure he possessed, yet, at the same time, he stripped his fields of practically all its other organic matter, gleaning every scrap of straw and stubble after reaping, for fuel. An evidence of the high standard of culture reached early in Chinese society has been the ability shown, even in the remote districts, to cope with the thorny problem of water use and distribution amongst the farmers, especially in the rice cropping areas. The water supply has always been a fruitful source of local quarrels, but on the whole, it has revealed a flair for co-operation. Nevertheless, the scientific use and proper conservancy of water has been lacking.

A further problem which confronted the farmer in the pre-1949 period, was that of under- and over-employment. In the farming year there were periods of great activity during planting and harvesting when every member of the peasant family was recruited and, if possible, labour hired. In the wheat growing regions of the North China Plain it is essential to take full advantage of the short periods of rainfall for ploughing and sowing. There are, however, periods in between, particularly in the long winter in the north, when there is marked under-employment. There are similar crests and troughs of activity in the rice-wheat growing areas of the Yangtze in winter and early spring, but in the far south, where there is an 11–12 month growing period, labour is fairly fully occupied throughout the year. The application of mechanical aids to meet labour demands in the peak periods and also the development of industry, which would absorb surplus farm labour, would help to solve this problem.

By far the greatest source of power used in agriculture has been unassisted human energy. A survey made by the University of Nanking in 1940 estimated that labour costs formed 33 per cent of the cost of production of rice, 41 per cent of wheat and 36 per cent of cotton. Animals supplement human labour on the farms to only a limited extent. Oxen, mules, donkeys and ponies—and even camels in Kansu—are used in the north, while in the south oxen and water-buffalo in the paddy fields are the important animals on the farm. Until recently no machinery has been used by the farmer. For carrying and haulage and for most of the heavy jobs on the farm he has had to rely on human labour.

It is on the reform of each and every one of the above faults and shortcomings that the present régime relies to achieve the phenomenal increases in production which it has set itself.

Land Use and Crop Distribution

In the following account of land use and of crops, care has been taken to use as few figures as possible. Most of the statistics used are those of the Department of Agricultural Economics of the National Agricultural Research Bureau of China, quoted by Dr. T. H. Shen in his *Agricultural Resources of China* and of Professor J. Lossing Buck's *Land Utilization in China.* While these are the most reliable data to be had of this pre-1949 period, they must be treated with reserve and must not be regarded as having the same accuracy as similar statistics of the UK or USA. Nevertheless they are very useful in giving a general picture of both production and distribution. The maps of crop distribution, appearing below, must serve for both the pre-1949 and post-1949 periods because no figures have yet been published by the present régime, which would enable new maps to be compiled. Although there have been big changes in production, the regional distribution will have remained, for the most part, undisturbed.

Mention has already been made more than once of the way in which China may be divided between north and south along the line of the Tsinling-Fu Niu Shan line. This is the appropriate place to emphasize how remarkable and striking a divide it is. Topographically it may be said to separate the mountainous north of the provinces of Shensi and Shansi from the hill-lands of the Yangtze and Si Kiang Basins, although there is a considerable amount of plainland common to both, where the North China Plain merges into the Hwai and then into the lower Yangtze flood plains. It is in this plain region that the distinction between north and south becomes somewhat blurred. Climatically the northern region is one of temperature extremes, dry and with a variable and inadequate rainfall and

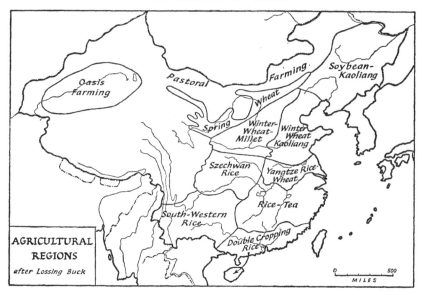

FIG. 34. Agricultural regions

with a short growing season. The south is more temperate, extending into sub-tropical and tropical realms, with a long growing period and much more reliable and plentiful rainfall. The soils of the two regions, as we have seen, stand in marked contrast, the north being unleached pedocals and the south highly leached and acid pedalfers. The lands north of the Tsinling are the homeland of the Han Jen and the population is more homogeneous in race than in the south. There are only two minorities of note: the Mongols of the northern borders and the Turki (Moslems) of Kansu. The Manchus are scattered and absorbed, and Manchuria is occupied mainly by Han Jen. As we have seen (p. 103), those south of the line are more heterogeneous. In the far south (Kwangtung and Kwangsi) they style themselves T'ang Jen and, while still essentially Chinese, they are distinct in stature, physiognomy, speech and temperament. Amongst them are many minorities.

Having all these contrasts in mind, it is not surprising to find that land use and crops of north and south are also very different. The only considerable pastoral farming of China is to be found in the north-west, notably in Kansu, Shansi, Suiyuan and Ninghsia. The rest of the agricultural land north of the Tsinling line is dry arable farm land; that is to say, it is cultivated without recourse to irrigation, other than wells, and is devoted to dry crops, such as wheat, kaoliang, millet, barley and cotton. In contrast, the characteristic crops of the south are rice and tea, the former being grown almost exclusively under irrigation. Very little upland or dry rice is grown. Wheat, barley and cotton also penetrate the northern borders of the south region to some extent. Even land ownership and tenancy are differentiated. Farms in the north are larger, and until 1949 there was a greater proportion of farmers owning their land than in the south. Furthermore, there was less reliance on human beings for power in the north, where there were more draught animals, more roads of a sort and more wheeled transport than in the south.

Lossing Buck in his *Land Utilization in China* divided the land into a number of agricultural regions, which are reproduced in Fig. 34. The regions are named according to the most characteristic crop or crops grown in the area. It will be noted that, with the exception of the Yangtze rice-wheat area, rice is the only cereal which figures south of the Tsinling-Fu Niu Shan line. It is estimated that there are approximately 340,000 sq. miles of cultivated land in China of which about 51 per cent lies north of this line in the 'wheat' lands and 49 per cent south in the 'rice' lands.

The crop production figures given below are averages over the years 1931–7, published by the Department of Agricultural Economics of the National Agricultural Research Bureau.

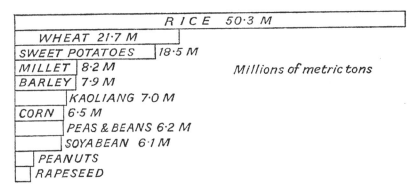

FIG. 35. Main food crop production in China Proper 1931–7

Food Crops

WHEAT

The dense wheat cropping areas are the North China Plain, i.e. Hopei, Honan and Shantung, the basins of the Wei and Fen, the middle Han Basin

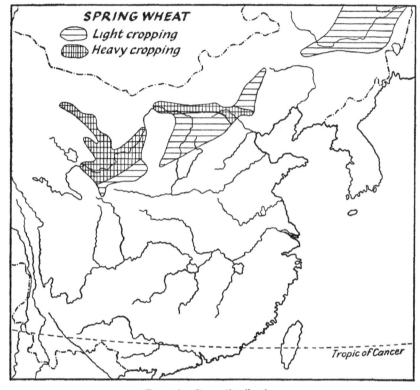

FIG. 36a. Crop distribution

in northern Hupeh and the river valleys of Szechwan. It is grown widely but not as intensively over the loess region of Shensi and Shansi. The annual production from 50 million acres was estimated to average 22 million metric tons between 1931 and 1937.

Beyond the Great Wall and beyond the 15 in. isohyet some spring wheat is grown, mainly in the 'pan-handle' of Kansu and Suiyuan. On account of the severe winters and the lack of winter moisture, planting is postponed till April and the crop is harvested in August. More careful seed selection could increase the yield in this area.

The great bulk of the crop is winter wheat. That grown in the wheat-millet and wheat-kaoliang areas is comparable to Canadian varieties in that it is hard, and is good for making bread and noodles. That grown in the Yangtze rice-wheat region is soft and good for confectionery and biscuits, rather similar to English wheat. There is a tendency, which is increasing, for more wheat to be grown at the expense of rice in this

PLATE VI: Rapeseed, which supplies an edible oil, being gathered in Kiangsu Province, 1960.

PLATE VII: The large state farms and collectives of the northern plains lend themselves to the use of machinery. This shows harvesting on the Chinchuan plain.

PLATE VIII: A production team reaping by sickle in the small terraced rice-fields. Here mechanization is much more difficult.

PLATE IX: Vegetables are grown everywhere in China, and are important in helping to balance the diet, which includes little meat and plenty of carbohydrates (wheat or rice).

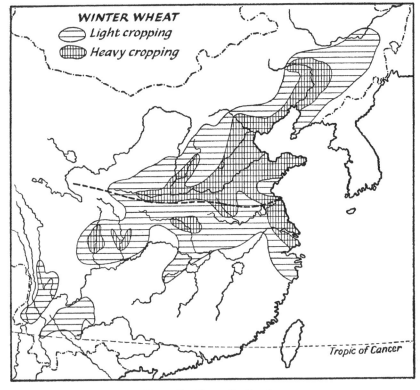

FIG. 36b. Crop distribution

southern region. Its food values are better balanced than rice, and especially in north Hupeh, where rainfall is more variable than farther south and the onset of summer rains less certain, the cultivation of wheat is less precarious. Wheat forms the most important cash crop of the north. According to Lossing Buck's survey, about three tenths of the crop was sold and the rest was consumed by the farm family.

Main Producing Areas (figures in million metric tons)

| Honan | 4·07 | Shantung | 3·58 | Kiangsu | 2·95 | Hopei | 1·8 |
| Szechwan | 1·8 | Anhwei | 1·4 | Hupeh | 1·2 | | |

BARLEY

This crop has a very wide distribution throughout China Proper with a main concentration in the Yangtze and Hwai valleys. It is grown largely as a winter crop in conjunction with rice. Barley stands up to wetter conditions

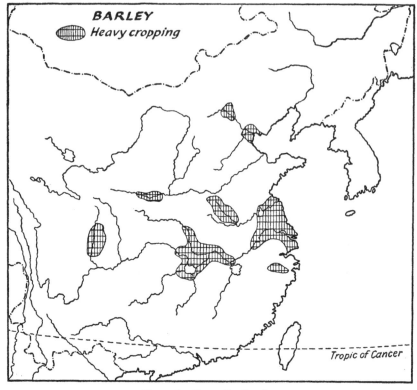

FIG. 36c. Crop distribution

better than wheat; it has a short growing period and matures early in May. This allows time for harvesting the barley and for ploughing the fields in readiness for the transplanting of the rice seedlings. Barley is used much more widely for human food than in Europe, although some is used for cattle feed and for brewing. Buck's estimate was 36 per cent human food; 32 per cent animal food; 18 per cent sold. Barley is also ploughed in as a green manure in the Yangtze valley rice fields. Estimated annual production was about 8 million tons or just over one third that of wheat.

Main Producing Areas (figures in million metric tons)

Szechwan 1·37   Kiangsu 1·35   Hupeh 1·03   Honan ·75

MILLET

This is essentially a northern crop which is grown in all the northern provinces. As in India, because of its drought-resisting qualities in this

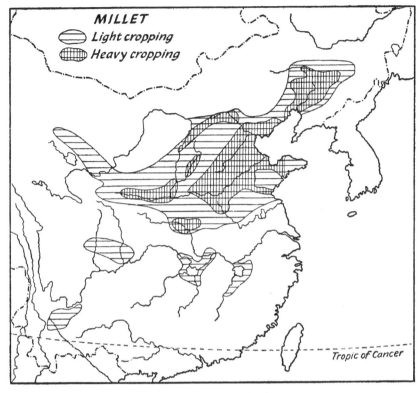

FIG. 36d. Crop distribution

region of variable rainfall and because of its short growing season, it is of great importance. It is the farmer's main insurance against famine. He usually grows wheat as a cash crop and keeps the millet, which he mixes with soy-bean flour and kaoliang, for home consumption. There are innumerable varieties, both glutinous and non-glutinous, the long fox-tail and barnyard being the most notable. Average annual production between 1931–7 was put at 8·2 million metric tons. Shantung, Hopei and Honan were the biggest producers, although wheat was a bigger crop both in weight and value. Although the actual production is not high as compared with the North China Plain, millet is also a most important crop in the loess lands of Shansi and Shensi.

Main Producing Areas (figures in million metric tons)

| Shantung | 1·84 | Honan | 1·16 | Shensi | ·22 |
| Hopei | 1·53 | Shansi | ·77 | | |

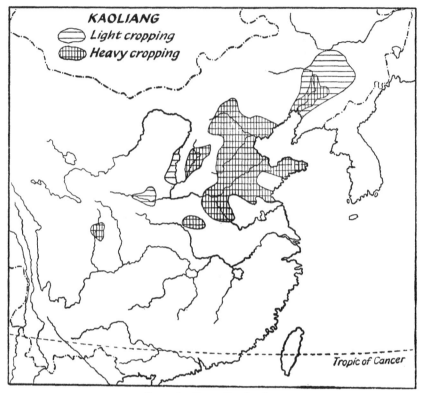

FIG. 36e. Crop distribution

KAOLIANG

This again is a northern crop. It shares the valuable drought-resisting qualities of the millets and is grown in conjunction with them. It is a tall plant with a strong stalk, 6 to 8 ft. in height, having a big, bushy head of grains, each about the size of a split pea. The grain, usually mixed with millet, is used largely for human food. Some is fed to livestock and some is used for wine-making. Very little figures as a cash crop. The stalk is valuable, being used for roofing, building, fencing and fuel. The threshed, bushy heads are used throughout northern and central China for brushes and brooms. Kaoliang is usually cultivated in widely spaced rows, and soy-beans, beans, peas and even wheat are planted between them.

Main Producing Areas (figures in million metric tons)

Manchuria (no figures available)

Shantung	1·93	Hopei	1·0	Shansi	0·47
Honan	1·14	Szechwan	0·6		

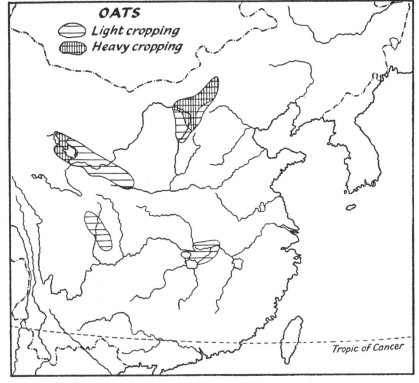

FIG. 36f. Crop distribution

OATS

Oats are not grown very much in China. The main region of production is on the higher lands of the spring wheat area where there is greater local precipitation. Oats are used mainly for human food.

CORN (MAIZE)

Corn or maize is a widely dispersed crop, being grown in a broad belt stretching north-eastward from Yunnan to Manchuria. Main production is centred in the Red Basin of Szechwan and on the North China Plain. Conditions for corn growing are excellent in the sub-tropical rice cropping areas, but little, in fact, is grown because of the intense competition of other crops, notably rice. Wherever it is grown, corn is used mainly for human consumption, and unlike the practice in the USA, very little is used to feed livestock. Both methods of cultivation and seed selection have been poor.

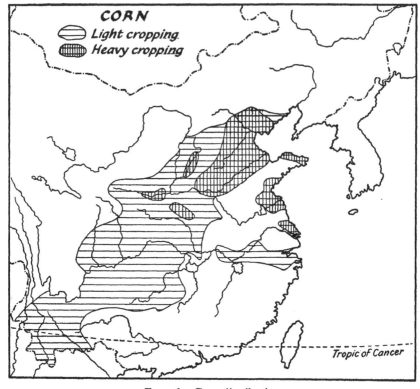

FIG. 36g. Crop distribution

T. H. Shen states that seed trials in 1947–8 showed that, using American hybrid varieties, the yield could be increased, in some cases, by as much as 200 per cent.[1]

Main Producing Areas (figures in million metric tons)

Szechwan	1·38	Shantung	0·80	Honan	0·65
Hopei	1·18	Kiangsu	0·65		

RICE

Although the dry cereals—barley, wheat and millet—were the earliest staple foods of the Chinese, rice has for long held the traditional place as the country's staple food. Despite the fact that wheat has better balanced nutritional values, rice is more popular and the most sought-after food.

[1] T. H. Shen, *Agricultural Resources of China* (New York 1951), p. 205.

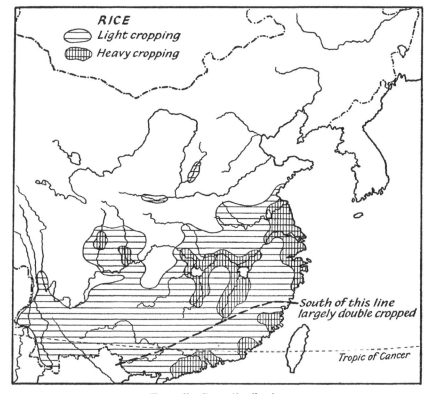

FIG. 36h. Crop distribution

It is grown wherever possible, and often in unsuitable conditions. For example, along the transitional zone between the rice-wheat and the wheat-kaoliang areas in northern Hupeh the farmers constantly waver between the attractive but dangerous lure of rice culture and the more solid but smaller cash returns of wheat.

The estimated area under rice is approximately the same as wheat, i.e. 50 million acres, but the yield is much greater. Wet rice produces some 2½ tons per acre as compared with just over 1 ton per acre by wheat. The conditions for its optimum production are a temperature of 21°C. (70°F.) or more over four consecutive months, during which there must be a rainfall adequate to keep the paddy fields flooded. A high relative humidity, which reduces evaporation, is helpful in retaining moisture. The soil must be such that it is able to retain the moisture. All these conditions are fulfilled in nearly all the low lands which lie south of the Tsinling line. Temperatures are high enough during the growing period, and with the onset of the summer monsoon, rainfall is nearly everywhere adequate. While the porous soils of the north are unable to retain the moisture near the surface, those of the south are heavier, and moreover tend to develop an impervious iron pan 18 to 24 in. below the surface, which is invaluable in preventing loss by seepage. One of the dangers of deep ploughing in southern China is that of breaking up this pan, and so allowing water to drain away.

Thus rice is produced in large quantities in all the provinces of the south. In the Yangtze valley it is possible to produce only one crop a year, the winter crop having to be devoted to wheat or barley. Passing southward over the Nan Ling, sub-tropical regions are reached in which double cropping is possible. In the tropical regions of south Kwangtung, Hainan and the New Territories of Hong Kong, where there is a twelve months growing period, a catch crop in addition to the two rice crops can be grown in December, January and February.

One further condition which is essential to high yield rice cultivation, at least until mechanization is introduced, is a plentiful supply of labour during the peak times, for, unlike Ceylon and Thailand, where much of the rice is sown by broadcasting, China cultivates the hard way. First the fields are ploughed, the baulks between the fields are repaired and the seed beds thickly sown. Then, as soon as the rains come and the fields are waterlogged, they are puddled; that is to say, cumbersome heavy wooden rakes are dragged backwards and forwards by water-buffalo until the whole surface has reached a consistency of thick potato soup. It is now ready for transplanting to take place. Bundles of seedlings are pulled up from the seed bed and tossed all over the field to be planted, and then a team of men and women laboriously plants them out in remarkably regular lines in stands of four to five seedlings together. From that time until shortly before harvesting the field must be kept thoroughly wet, if possible continually with a covering of four to five inches of water. Where this is possible the field may be used simultaneously as a fish rearing pond, which has the additional advantage of keeping the water free from mosquito larvae. If a good crop is to be secured, the field must be weeded two or three times before harvest. Reaping is done by hand, usually by means of toothed sickles and threshing done in the field by beating the heads of grain over the edge of a screened tub. The straw is used for fuel and for animal feed. Very little upland, i.e. non-irrigated, rice is grown in China.

In spite of the fact that China is so great a rice producer, it was unable before 1949 to meet all its needs. Szechwan and Hunan were the only provinces which produced a sufficient surplus to be able to supply other parts of the country to any appreciable amount. Rice was imported from Indo-China, Thailand and Burma into Canton, Swatow Amoy, Shanghai and Tientsin. The amount imported varied from year to year very considerably, according to the rice harvest in China. Between 1933 and 1937 it ranged from 310,349 tons to 1,296,448 tons and averaged 803,795 tons per annum.

Main Producing Areas (figures in million metric tons)

Kwangtung	7·83	Kwangsi	3·34
Szechwan	7·13	Kiangsi	3·26
Hunan	4·86	Anhwei	2·16
Kiangsu	4·21	Fukien	2·15
Chekiang	3·91	Yunnan	1·61
Hupeh	3·51	Kweichow	1·08

VEGETABLES

Vegetables are grown everywhere in China and are of immense importance. They help to balance what is usually a very poorly balanced diet, which includes very little meat and is overloaded with carbohydrates, (wheat or rice).

On account of the size of the country there is a great variety of vegetables produced, and many districts are famous for some particular kind, e.g. Hupeh for its *hong ts'ai* or red cabbage. There is one vegetable which is ubiquitous—*pei ts'ai* or white cabbage—which makes its appearance at most meals. For the rest—turnips, onions, garlic, radishes, cucumbers, tomatoes, string beans, peas, melons and squashes are grown, with increasing emphasis on melons and squashes in the south.

Around the big cities intensive market gardening is carried on. This is increasing with the improvement of local communications and the use of mechanical transport in the vicinity.

SWEET POTATO

The sweet potato was introduced into China via Fukien from the Philippines in 1594, and from there it has spread all over the country to the great advantage of the poor farmer since it forms a big item in his diet. It is especially valuable in poor years of drought, for the vines and leaves form a soil cover which prevents loss of moisture by evaporation. The vines are fed to the pigs.

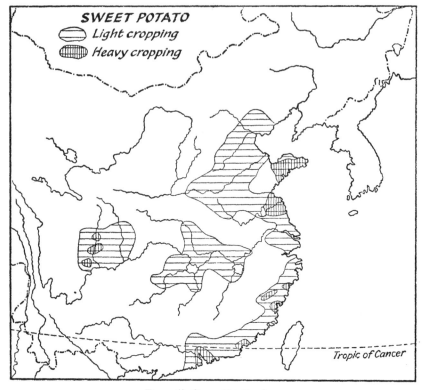

FIG. 36i. Crop distribution

The sweet potato is sown in June and harvested in November, and produces a heavy crop. Its harvesting calls for some judgement. If taken too early, the potato contains too much moisture, and if too late, it is caught by the frost. In neither case will it store well.

Main Producing Areas (figures in million metric tons)

Szechwan	2·58	Kiangsu	1·94
Honan	2·29	Hopei	1·53
Shantung	2·28	Fukien	1·12
Kwangtung	1·97		

FRUIT

The size of the country and the resulting great variety in climate produce an even greater range of fruit than of vegetables. In the north and centre temperate fruits, apples, pears, peaches and apricots thrive. In sub-tropical parts mandarin oranges, persimmons, pomelo and papaya are grown, while in the tropical south of Hainan and southern Kwangtung pineapple, banana and mango can be cultivated.

Although the whole country is so full of promise, Chinese agriculturalists have been singularly blind to the possibilities of fruit growing. It has generally been regarded by the farmer as of very secondary importance, worthy only of the casual and haphazard planting of a few trees here and there on waste or unwanted land. The trees have been left to fend for themselves and even to suffer the indignity of being lopped for firewood.

Quite recently and suddenly in the last three decades, the Chinese have become very fruit conscious. Thirty years ago it would have been surprising to find a fruit stall of any size other than in the big cities which were accessible to overseas trade. Today the picture has changed and it is a poor village that does not have its fruit stall. This is an industry which is developing very rapidly and in which there is ample scope for improvement in all its departments, in cultural methods, in the orchards themselves, in the selection of the right varieties, in marketing and in the industrial use of the fruit (for canning, drying, etc.).

SUGAR

As with fruit, so with sugar. Mainland China has never given serious attention to the production of sugar. In Kwangtung, around the delta of the Si Kiang and in Hainan, individual farmers have devoted small patches of their land to the cultivation of sugar cane, but it is nearly always regarded as of secondary importance. It is small wonder that what is grown is often of poor quality. Nor is it surprising to learn that the Chinese rank very low in sugar consumption per capita. It is estimated that only 6 lb. per head is consumed per annum as compared with about 128 lb. per head in Australia. Practically all sugar produced on the mainland is processed by old fashioned millstone extraction methods, which are very inefficient and wasteful.

In striking contrast is the sugar industry of Taiwan, which was developed and fostered by the Japanese after their annexation of the island in 1896. Here the sugar farms are often large enough to rank as plantations and are using modern methods of cultivation, transport and processing. Taiwan ranks third in the world's sugar-cane production.

Main Producing Areas (figures in million metric tons)

Taiwan	0·938	Szechwan	0·10
Kwangtung	0·13	Fukien	0·05

VEGETABLE OILS

Vegetable oils play a big part in the diet of most Chinese. As they eat very little meat and, except in the north-west, drink even less milk, these oils form their main source of fats. They are used universally in cooking, and until the introduction of kerosene, provided the main source of lighting. They have figured as one of China's chief exports.

The oils can be divided into two general classes, edible and industrial, but the classification is not mutually exclusive. Some oils fall into both categories.

The chief edible oils are soy-bean, peanut, rapeseed, sesamum and cotton seed, and of these, the soy-bean is pre-eminent. It is grown in every province but is essentially a temperate plant. Before the Second World War, Manchuria produced an average of 4·27 million tons per annum, Shantung 1·57 million, Kiangsu 1·13 million and Honan 0·73 million. Until the outbreak of war, Manchuria, under the Japanese, had a virtual monopoly of world trade in both soy-seed and oil. The USA started production during the war, and now at least equals, if it does not surpass, that of China. Manchuria's export has fallen to less than half.

The soy-bean has many uses. It is a very important food: bean curd (*tou fu*) appears at practically every meal in humble households and it is served in literally a hundred different forms. The bean is crushed for oil, which is used in cooking as butter and lard substitute, and in paint and soap making. The residue is pressed into large round cakes for cattle feed and for making explosives. It was in great demand during the First World War. Last, but not least, its growth has helped to secure, through the centuries, adequate nitrogen content in the soil.

Before 1949 peanut production was second only to India and averaged 2·73 million tons (Shantung 0·63; Hopei 0·42; Kiangsu 0·32). Rape seed averaged 2·47 million tons (Szechwan 0·62; Kiangsi 0·28; Hunan 0·24; Chekiang 0·20). Sesamum averaged 0·85 million tons (Honan 0·23; Hopei 0·10; Shantung 0·10). Cotton Seed averaged 1·61 million tons (Kiangsu 0·36; Hopei 0·30; Shantung 0·20; Hupeh 0·18).

The main industrial oils are *t'ung yu* (wood oil), hemp, tallow, linseed, tea and castor. Of these *t'ung yu* is by far the most important. It has a wide distribution in those provinces south of Tsinling, with a marked concentration in the Yangtze valley. The oil is derived from the nut of a tree and its production evinces yet again the Chinese farmer's surprising carelessness in regard to tree culture, which has amounted almost to contempt. This is all the more surprising in view of the high regard in which trees are held in Chinese art. *T'ung yu* trees are often carelessly planted too close together and frequently the lower branches are lopped for firewood. *T'ung yu* is produced both for local use and for export. It is a basis for paints and varnish and is used as a preservative on most Yangtze river craft; also in making *yu pu* (oiled cloth) which is widely used throughout the country. It was formerly used as a lighting fuel. It is noteworthy that it was the sudden growth of the *t'ung yu* export trade which saved Hankow from economic disaster when its tea trade with Europe collapsed in the 1880's.

Flax and hemp are grown both for their fibre and their seed. Both linseed and hempseed have a high oil content of between 30 and 40 per cent. These oils are important in China for cooking purposes and for export. They used to be sent mainly to Japan, for lubricating oil. The residue, after crushing, is used for a fertilizer. The main growing region for flax and hemp is north of the Tsinling line, flax being concentrated in the area north of Peking, in the Fen valley and in Kansu, and hemp in northern Manchuria. Most of the castor seed, vegetable tallow and tea seed is grown south of the Tsinling line in the Yangtze valley provinces.

TEA

Mention is made in very early Chinese writings of the use of tea as a medicine. By AD 550 it was in use in China as a beverage and by the eighth century its use had become so common that it was a subject of taxation. Since then it has been the national beverage. With the exception of wine at feasts, tea and boiled water are virtually the only drinks throughout the length and breadth of the land. It is interesting to speculate, in terms of human lives saved, how much the simple fact of boiling water in order to make tea has meant. China has suffered from flood and famine but not from the sweeping scourge of cholera.

Four main kinds of tea are produced: black, green, brick and scented. The difference between black and green tea lies not in the leaf but in the fact that the black leaf is fermented and the green is not. Green tea is fired at once after picking, and so fermentation is prevented. This gives a softer, more delicate brew, which is appreciated throughout China and is popular in both north-west Africa and in the USA. Black tea goes through the whole process of withering, rolling, fermenting and firing. Brick tea can be either green or black. The term simply describes the form in which the tea is prepared for transport. It is pressed into tiles, about 5 in. square and ½ in. thick, rather than bricks. In this form tea has been exported for centuries on the backs of coolies from Central China into Tibet, Mongolia and Russia. Scented tea is prepared by layering either green or black tea with chrysanthemum, jasmine, rose or gardenia. A fifth kind of tea, known as *oolong*, differs from green and black in being semi-fermented.

Black tea is grown widely over the whole of the tea-producing area. Green tea is concentrated mainly in Anhwei, Chekiang, Kiangsi and Taiwan, while brick tea is produced mainly in Hupeh and Hunan. The total production of tea in China is estimated to be approximately 400,000 tons per annum.

In 1557 the Portuguese established their trading station at Macao, and through trade and diplomatic relations acquired a knowledge of tea, and introduced it to Europe. The first tea house in London was established in 1657. Pepys, writing in his diary in 1660, says 'I sent for a cup of tay, a China Drink, which I had never drank before', but it did not become a popular drink in Great Britain until Indian tea was firmly on the market at the end of the nineteenth century.

Very little tea came out of China until trade with Europe was opened after the First and Second China Wars of 1840 and 1856. In 1838 only eight small chests were exported, but after 1860, when the Yangtze was opened, the trade grew rapidly. This was the period of the famous China clippers, fast sailing vessels which raced to London each year with their tea cargoes. The first home got the best prices. Until 1880 China had a monopoly of the tea trade of the world, when it was exporting 300 million lb. annually. Then Indian tea quickly began to make its appearance and within a decade Chinese tea had completely lost control of the market, which it has never since recovered. Even in 1935 its tea exports amounted to only 84 million lb. The reasons for this loss are many. Probably the chief reason is that Indian and Ceylon teas make a darker brew with more 'bite' than China tea. The stronger tea is more popular in Great Britain and Ireland, the chief consumers. Indian and Ceylon teas are grown on a large scale on big plantations, using up-to-date machinery and scientific methods. In China tea was grown almost entirely by small peasant farmers as a secondary crop to their rice. It received only haphazard attention, was cultivated by poor methods, picked carelessly and was slipshod and irregular in its grading and packing. Lastly, the export trade in tea was subjected to heavy Chinese taxation.

Livestock

PIGS

Pork is the most popular meat and is eaten when it can be afforded by all except those who avoid it for religious reasons. In fact, the pig is the only animal which is widely consumed in the country. There are no big pig farms but every farming household has at least one sow. The pig is particularly well suited to the rural economic conditions where the poor peasant is unable to buy fodder for livestock. The pig is a scavenger and is largely left to root for itself. It is given such waste products from household food as there are, together with sweet potato vines and the like, but apart from these it has to fend for itself.

Bristles form a valuable by-product and they are still exported in considerable quantities, although, with the rise of synthetic products, the demand for bristles has dwindled. Pigs are also considered valuable on the farms for their manure.

The estimated annual average number of pigs in China between 1931–7 was 66·34 million, yet the estimated amount of pork eaten per person per annum was only about 20 lb. The following figures (in millions) of the distribution of pigs throughout the country reveal how ubiquitous the pig is.

Szechwan	8·18	Kwangsi	3·65
Kwangtung	5·19	Shantung	3·51
Hunan	5·03	Honan	3·19
Kiangsu	5·02	Anhwei	2·80
Hupeh	3·93	Yunnan	2·76
Kiangsi	3·80	Chekiang	2·72
Hopei	3·74		

CATTLE

Cattle are used in China almost exclusively for draught purposes. Oxen are used in both north and south for work in the dry fields, and water-buffaloes are used in the wet rice paddy. It is estimated that there are 11·34 million water-buffalo south of the Tsinling line as compared with only 0·24 million north of the line. Beef figures very rarely in Chinese meals. The fact is that a much greater return in food values per acre can be obtained by arable farming, producing rice and wheat, than by raising cattle. Pressure on the land does not permit beef cattle raising in the heavily populated parts of China. It may be that when the natural grasslands of the north-west are systematically and scientifically developed for pastoral farming, and when railway communications are adequate for large scale refrigerator transport, beef will become a more popular item in Chinese diet.

Until very recently milk has been eschewed by the Chinese, but now its virtues are being recognized and the demand for it is increasing rapidly in the big cities around which small herds of good diary cattle are kept. The milk yield generally is very low as is to be expected when the animals are kept primarily for draught purposes. The water buffalo is more prolific in its yield, but unfortunately its milk is not acceptable either for colour or taste. It is interesting to note that Chinese babies are weaned, not on cow's milk, but on *tou chiang*, soy-bean curd, which is rich in vitamins.

Main Cattle Areas (figures in millions of head of cattle)

Honan	3·14	Hunan	1·37
Shantung	2·58	Sinkiang	1·21
Hupeh	1·87	Hopei	1·17
Kiangsi	1·74		

Main Water Buffalo Areas (figures in millions)

Szechwan	2·00	Hupeh	0·92
Hunan	1·48	Kiangsu	0·91
Kwangsi	1·38	Kiangsi	0·83
Kwangtung	1·30		

SHEEP

Mutton is a meat which arouses little enthusiasm amongst the Chinese generally and evokes active dislike widely in the centre and south. Sheep rearing is confined almost exclusively to the north-west as the following figures (in millions) demonstrate: Sinkiang 9·43 (probably including goats); Kansu 2·62; Shansi 2·20; Suiyuan 1·41; Shantung 1·09; Honan 0·89. It is estimated that there are less than 20,000 sheep south of Nan Ling, and that even in the Yunnan-Kweichow plateau there are only some 200,000.

POULTRY

The main chicken population of China, which is very large, is found south of the Tsinling line. Chicken are reared on every farm. Scant attention is given to selection of type and little care to feeding, with a result that the toll from disease is very severe, and the yield of both flesh and eggs is a great deal less than it could be. Nevertheless, chicken is second only to pork in meat diet, and since the development of international trade, particularly in this century, eggs (dried and frozen) have been an important item. Before the Second World War Hankow processed 2 million eggs per day. Chicken feathers are used for filler and for Chinese 'pens', i.e. brushes for calligraphy.

Main Chicken Areas (figures in millions of birds)

Kwangtung	22·06	Kiangsi	17·53
Hupeh	21·46	Kwangsi	16·97
Shantung	21·35	Chekiang	16·44
Honan	18·37	Szechwan	15·85
Kiangsu	18·39	Anhwei	15·83

DUCKS

Duck, while not as numerous as chicken, is an important item of food. As might be expected, ducks are reared mainly south of the Tsinling line in the wetter lands of the Yangtze valley and in Kwangtung and Kwangsi. There are two main kinds: the Peking duck and the Nanking duck. The former is used for roasting and is an important dish at all feasts. The latter is salted and widely distributed. Perhaps a word should be said of the fabulous Chinese preserved or 'Ming' eggs, reputed to be many years old. This, of course, is not so. They are ducks' eggs, preserved for some forty days in lime, during which time they turn black. They are much appreciated everywhere in China. When the foreigner overcomes his repugnance to the colour, he finds them delicate in flavour.

Main Rearing Areas (figures in millions)

Kwangtung	7·42	Szechwan	5·30
Hunan	6·57	Kiangsi	4·68
Kwangsi	6·34	Anhwei	3·57
Kiangsu	6·19	Hupeh	2·48

FISH

While the Chinese are not so dependent on fish for essential proteins as the Japanese, they nevertheless eat large quantities of fish, which are very valuable in helping to attain a balanced diet.

There are two main sources—marine and fresh water—both of which are important. A wide continental shelf underlying the Hwang Hai (Yellow Sea), the Tung Hai (East Sea) and the Nan Hai (South Sea), stretches the whole length of the east coast. In the north is the Gulf of Pei Chihli, which is largely enclosed by the Shantung and Liaotung peninsulas and into which the waters of the Hwang Ho flow. All these waters abound in fish.

Practically all the sea fishing is done by sailing junks, which numbered about 63,000 in 1937. At that time there was very little mechanization, there being only 771 motorized vessels, half of which were operated by Japanese, who were and still are, pre-eminent in the fishing industry of the Far East. Off-shore fishing is done by long line, trawl and drift net. The chief species caught in the north and central seas is the yellow croaker (*hwang hua*) and in the south, the golden thread (*hung hsien*). Little attempt has yet been made to share in the rich fishing grounds farther afield, such as the tuna to the south of Taiwan or the salmon of the cold seas of the north. This development cannot come, however, until there has been a big extension of communications and of refrigeration. At present catches must be consumed near the port of landing.

Fresh water fisheries are more important than marine in that they are developed over a much wider area and serve a greater number of people. All the many large rivers are teeming with fish, and the lake-studded plain of the lower Yangtze is a great fishing ground. Nearly every village in the whole country has its pond, which is sited just below the houses and is the receptacle of both vegetable and human waste, which provides food for the fish fry with which it is stocked every spring.

There has been a considerable development of pisciculture, particularly in south China in recent decades, in artificial ponds on a relatively large commercial scale. The main species stocked in the south are carp and mullet. Pig, cow and poultry dung and night soil are the main foods fed to the fish in these ponds.[1] The average return per *mow* is double that of paddy, but the initial capital outlay and cost of stocking are considerable.

Industrial Crops

COTTON

Tradition has it that Marco Polo first introduced cotton seed into China, and that the emperor wore the first robe woven from it. From another source[2] we have it that cotton first came from Khotan in the eleventh century and was introduced only in the face of strenuous opposition by the silk growers. Whatever the origin, it is certain that during the ensuing six centuries only small quantities were cultivated. In the closing years of the nineteenth century and with the introduction of American varieties, cotton production grew rapidly. In 1894 the first modern cotton mill was opened in Shanghai.

---

[1] T. R. Tregear, *Land Use in Hong Kong and the New Territories* (Hong Kong 1958), pp. 65–68.

[2] E. H. Wilson, *A Naturalist in West China* (London 1913), Vol. 2, p. 83.

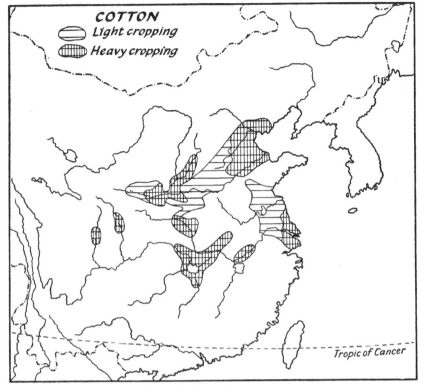

FIG. 36j. Crop distribution

There are two important cotton belts, one to the north of Tsinling and the other to the south. Each produces about half the total crop.

In the northern belt rainfall is not adequate, being only 20 in. in many parts and not arriving until late spring. Therefore irrigation is necessary. The growing season, too, is short (180 days) and so early maturing varieties are needed. For this reason Chinese cotton has been preferred as it required three to four weeks shorter growing period than American varieties, and so could be grown more easily in rotation with winter wheat. More recently American Upland (Stoneville) and King's Upland, because of their short growing periods, have gained favour on the North China Plain and in the lower Liao valley respectively.

Rainfall in the southern belt is heavier and falls mainly in the growing period. No irrigation is necessary. The growing period is longer and there is little or no danger of frost. Here Chinese varieties are generally preferred to American as they are more resistant to pests and because the boll droops, in contrast to the American, which stands up. Thus it is better adapted to the greater humidity of the Yangtze Valley.

All Chinese varieties are short staple, although they are graded 'long' and 'short'. The long staple is only 0·7–0·8 in. in length. Nevertheless, some varieties have a high yield of lint: *hsi ho* yield 35 per cent lint as compared with mixed American-Chinese giving only 30 per cent.

A little long staple Sea Island and Egyptian cotton is grown in the south in Kwangtung, Hainan and Taiwan, but conditions are not favourable to its production for the rainfall is too heavy.

It was estimated that between 1931–7 the average production was 809,500 metric tons of lint. The Chinese are a cotton wearing people, and this amount was inadequate to meet the needs of even a very low standard of clothing. During this same period there was an average import of 84,218 metric tons. After the war imports increased considerably owing to the fall of production during Japanese occupation. In 1947 cotton imports rose to 197,000 tons.

As with so many other crops, cotton growing has suffered from poor methods of cultivation arising out of ignorance, poverty and insecurity. Much of the cotton is grown on a small scale by peasant farmers. Better selection of seed and type, more adequate application of chemical fertilizers and a country-wide attack on cotton diseases and pests would result in a yield probably big enough to make the country self-sufficient. Happily China's cotton has not yet fallen a victim to the boll weevil.

Main Producing Areas (figures in thousands of metric tons)

Kiangsu	177·5	Shensi	43·0
Shantung	92·5	Szechwan	34·5
Hupeh	91·0	Shansi	32·5
Honan	88·0		

SILK

Tradition has it that the knowledge of sericulture dates from 2,700 BC when Hsu Lung-she discovered the art of reeling and weaving silk. Although the knowledge of the art was widespread in Chou and Han times, the secret of silkworm rearing was kept within China until AD 419, when, according to a pretty legend, a Chinese princess, betrothed to the King of Khotan, smuggled silkworm eggs and the seeds of the mulberry tree in her trousseau. Sericulture in Japan dates from AD 794.

We have seen (pp. 82–86) something of the early silk trade with

Rome and the routes taken. Although that trade declined, silk continued through the centuries to be an important commodity of internal commerce.

All silk producing caterpillars belong to the *Bombyx* family of which the outstanding species is the *Bombyx mori*. The 'worms' are reared from eggs, which are hatched at a temperature of about 75°F. and take from 40 to 45 days from hatching to the spinning of the cocoon. Their food is the mulberry leaf, which they need in great abundance, for it requires about 1,000 lb. of leaves, which is the crop of 25 to 30 trees, to rear one ounce of eggs, numbering 35,000. Since the necessary temperature requirements can now be produced artificially, the climate that will give an adequate supply of mulberry leaves is more important than that needed for incubating and rearing the silkworms.

The main producing areas are around the estuary of the Yangtze in Chekiang and Kiangsu, where the famous China or white silk is reared; in Kwangtung around the Pearl River delta, where the silk is noted for its softness; and in Szechwan, where yellow silks are produced. A relatively small amount of tussah comes from Shantung.

Some stimulus was given to the moribund silk export trade at the turn of the century by a progressive viceroy, Chang Chih-tung, who made real efforts to instruct the peasantry in modern methods and also to improve marketing. Nearly all sericulture in China is carried on as a cottage industry, supplementary to rice. In Chekiang and Kiangsu, where there is only one rice crop a year, sericulture competes very keenly for the available labour between April and August.

Here again production is marred by haphazard methods and ignorance. In the homesteads where most of the rearing is done, disease, once it has started, is likely to continue through constant re-infection. The proper and essential control of temperature and humidity is very difficult to achieve in the given conditions; consequently, losses up to 70 per cent are not uncommon. Most of the filaturing and reeling is done in the cottages with the crudest tools with the result that it is difficult to standardize quality. Some advance was made before 1949 through co-operative societies[1], but the ravages of the Sino-Japanese war, together with the increasing competition by synthetic products and also the far more efficient Japanese sericulture, have combined to deal Chinese silk production a heavy blow. Total production from the main areas in 1937 was estimated at 205,000 bales and in 1947 at only 62,000 bales.

[1] Fei Hsiao-tung, *Peasant Life in China* (London 1947), Chapter 12, discusses fully the silk industry in Chekiang.

Main Producing Areas (figures in metric tons)

Chekiang	4,450	Hupeh	460
Kwangtung	3,325	Shantung	375
Szechwan	1,750	Anhwei	285
Kiangsu	1,525	Honan	165

FIBRES

Ramie (*Ssu ma*—silky flax). This, the most important of Chinese fibres, is a perennial plant grown entirely in the Yangtze basin, the main production being in Hunan, Kiangsi, Hupeh and Szechwan, in that order. It flourishes on well drained, warm hillsides. The ramie plant can be cropped for about twenty years but its best cropping time is in the seventh and eighth years. The plant grows to a height of between 3 to 6 ft. and produces long stalks from which strong and fine fibre is obtained by retting. Three or four harvests a year can be reaped. From the fibre the famous grass cloth, the Chinese equivalent of Irish linen, is woven. Grass cloth is popular with all classes in the summer, since in the hot, humid climate of the centre and south, it allows a freer passage of air to the skin than does cotton cloth. It is also used for strong cordage, strong fabrics and fishing nets.

Hemp (*Ho ma*). This is an annual plant which is cultivated mainly in Manchuria and north China. It is planted in February and harvested at the end of May, growing to about 8 ft. in height. The hemp is retted and sun dried, and the fibre separated from the bark and woody stem. This woody residue is burned and the ashes used in the manufacture of firecrackers.

Hemp fibre is used for cordage and for coarse summer cloth. The average annual production is about 90,000 metric tons.

Palm Fibre. This is the bract of coarse fibre encasing the stem of the palm. As cut from the tree it forms pieces of natural fibre cloth which is usually used in that state by the peasants for rain cloaks. These are especially useful for rice planting and work in the fields in the wet summer.

Tobacco

Tobacco was probably introduced into China about the same time as corn, i.e. approximately 1530. Since that time it has increased steadily in use for smoking, for snuff and medicinally, in spite of occasional official opposition. Until recently only the pipe and water-pipe were smoked. This is a significant fact because the Chinese pipe uses very little tobacco—a

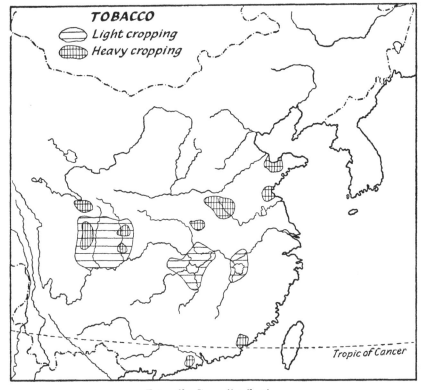

FIG. 36k. Crop distribution

pinch of tobacco, ignited by means of a slow match and sufficient for no more than three or four puffs before refilling. Thus locally grown tobacco in those days was sufficient for local needs. It was almost entirely sun cured.

Since the First World War the habit of cigarette smoking has grown enormously. This has led to a great increase in the growth of tobacco similar to American Virginian and to flue-curing, rising from 1,089 metric tons in 1916 to about 95,000 in 1937 and 113,399 in 1947.[1]

Main Producing Areas 1947 (figures in metric tons)

Honan	39,952	Szechwan	8,231
Anhwei	15,000	Yunnan	4,148
Shantung	13,608	Kweichow	3,500

Manchuria produced 16,230 metric tons in 1946.

[1] T. H. Shen, op. cit., p. 280.

Industrial Development

We have already seen that the merchant stood low in the Confucian order of social classes. It was clearly to the interest of the scholar-landlords that no other class arose to challenge their superiority, and this they successfully did until the beginning of the twentieth century by discouraging the development of manufacturing industry beyond the handicraft stage.

Until the middle of the eighteenth century Chinese and Western industrial organization were very similar. Manufacture of all kinds was in the hands of merchants and small masters, in small units, employing a mere handful of workmen who used primitive tools and no mechanical power. If there were differences, they consisted in the greater diversity of Chinese crafts and in the general superiority of Chinese craftsmanship. While the West in the next 150 years experienced all those fundamental changes covered by the term *Industrial Revolution*, Chinese methods and organization remained virtually unaffected until the turn of this century.

There were several reasons why China remained so long untouched by the industrial changes taking place in the West, but most of them stem from the conviction, held throughout the centuries, of Chinese superiority and self sufficiency, epitomized in the very name they gave their country, 'Chung Kuo' or Central Kingdom. This 'chosen race' concept inevitably carried with it both complacency and isolationism. When Lord Macartney led a mission from George III in 1793 to the great Emperor Ch'ien Lung with propositions of mutual trade, he was met with the retort 'Our Celestial Empire possesses all things in prolific abundance and lacks no product within its own borders; there is therefore no need to import the manufactures of outside barbarians.'[1]

While Western powers, notably Great Britain, hammered at the closed Chinese door, and as a result of two wars in 1840–42 and 1856–60 forced an entry, China remained to all intents and purposes industrially undisturbed until there arose in the last decade of the nineteenth century a court progressive party, which, with the Emperor Kuang Hsu's backing, endeavoured to introduce political and economic reforms of a Western pattern. The attempt met with violent opposition from the conservative element, led by the Dowager Empress, Tz'u Hsi, who inspired the Boxer Uprising, which was really an attempt to oust the foreigner and all his ways. In 1897 the Dowager Empress seized Kuang Hsu and encouraged the reactionaries to perpetrate atrocities on, and to massacre foreigners and Chinese Chris-

[1] Sir Edmund T. Backhouse and J. O. Bland, *Annals of the Court of Peking* (Peking 1939), p. 326.

tians. The result was that the Western powers intervened militarily, Peking fell, the court fled, Russia took possession of most of Manchuria and a huge indemnity was imposed, which eventually was devoted almost exclusively to the education of Chinese students in overseas universities and technical institutions, and so had far-reaching effects. Then followed a long period of political unrest, revolution and chaos. The Russo-Japanese War (1904-5) left Japan paramount in Korea and Manchuria, a cockpit of international strife. In 1911 the Manchu dynasty was overthrown and a republic declared, only to be followed by civil war between contending warlords until a second revolution in 1926-7 established the Kuomintang.

The significance of this period of unrest from the point of view of industrial development is that it was during this chaos that real beginnings of change from medieval craft or cottage industry to modern industrialization took place. Foreign capital began to flow more freely into the country, and Chinese capital began to be mobilized along Western lines. Railways were constructed, mining was developed and modern mills were built. But modern development, because of its early reliance on foreign initiative and capital, and because of poor internal communications, was mainly along the coast and up the navigable Yangtze in the treaty ports.

It is easy to exaggerate the extent of industrialization that occurred between 1900 and 1930. In actual amount it scarcely touched the vast country. At this time big cities presented a most interesting exhibition of all stages of industrial development working literally side by side. The writer remembers in a single walk through the streets of Wuchang in 1925 or 1926 seeing first a man at work in a dark corner of a cottage on a single hand loom. Not much farther down the street was a workshop in which were collected about a dozen similar handlooms, with workers operating under the eye of a foreman; while, not far away, down at the riverside, was the huge No. 1 Cotton Mill with the then up-to-date machinery from Lancashire and employing hundreds of workers.

The few years immediately preceding the Japanese invasion in 1937 were marked by signs of real industrial progress. The British Commercial Counsellor at Shanghai, commenting at that time, said 'The outstanding feature of Chinese economic life is the increasing justified confidence which the Chinese themselves as well as the world at large have in the future of this country, a confidence based on the remarkable growth of stability achieved in recent years, and the improved political, financial and economic conduct of affairs—government and private.'[1] It may well be that it was

[1] A. Donnithorne, 'Economic Development in China', *The World Today*, Vol. 17, No. 4, 1961.

this growing stability and industrial strength which led Japan to attack China in 1937. Be that as it may, it is certain that China emerged in 1946, after long years of war, weakened, divided and rent by civil war, floundering in inflation and administrative corruption—chaos in which further economic development was impossible.

Apart from the general requirements of peace, honesty in public affairs, stable currency, what were China's known natural resources for modern industrialization and how have they been developed?

Mineral Resources[1]

We have seen (Chapter 2) that there was very early a knowledge of metallurgy in China. In Shang times magnificent bronze sacrificial vessels were cast, the bronze itself probably coming from the south (Chu) and later from Yunnan. Iron was scarce and in very short supply. It was not until the Later Chou that it began to oust bronze. Copper 'cash' came into use as currency at this time and was still in circulation throughout the country in the 1920's. It was suspended for a while in the Later Han in favour of silk and grain, for there never was enough copper adequately to meet the currency needs of the country. The fact is that mining was held in very low esteem: miners were classed with soldiers, and thieves as the lowest of the low, it being held that miners stole from the earth in contrast to the peasant. Moreover, mining interfered with *feng shui*, the spirit of wind and water. There was, and still is, an animistic belief that powerful spirits (dragons, phoenix, tigers, stags and turtles) reside in all natural phenomena, the hills, rocks, water, air and trees. The geomancers, who studied these things and dealt in the art, were powerful persons. It was they who decided where and when burial should take place, care being taken that the lowly born were buried at lower levels than the princely. It was they who decided whether any action involving the use of the earth for mining, building, etc. would disturb or offend the *feng shui*. This militated against mineral exploitation throughout the centuries, but never so severely as in the period from about 1880 onwards, when Western ideas and innovations were making themselves felt. There was endless obstruction to new mining enterprises; railway construction was everywhere resisted; telegraph poles were cut down. The overcoming of this superstitious opposition has been long and has been a real factor in slowing

---

[1] See W. F. Collins, *Mineral Enterprise* (London 1918), for an excellent account of an early mineral development.

down economic development. Nor has it yet been entirely overcome even in so advanced an area as Hong Kong. Only recently *feng shui* was invoked by the local peasantry to prevent the Hong Kong government from building a much needed road along its natural route on the island of Lantao.

COAL

Marco Polo wrote that 'it is a fact that all over the country of Cathay there is a kind of black stone existing in beds in the mountains, which they dig out and burn like firewood'. However, it was Baron Ferdinand von Richthofen, who, between 1870 and 1872, first revealed to the West the vastness of China's great mineral wealth and particularly her great reserves of coal. Her richest fields are in the north, but coal is widespread throughout the country and is to be found in every province. Once again we find the Tsinling line as a divide between north and south; this time as a geological divide. The coal beds north of the line are essentially Palaezoic in origin, while those to the south are mainly Mesozoic.

Northern China has not been deeply submerged since Ordovician times. During the Carboniferous and Permian eras it was covered by a shallow sea and broad continental shelf, which oscillated between land and sea,

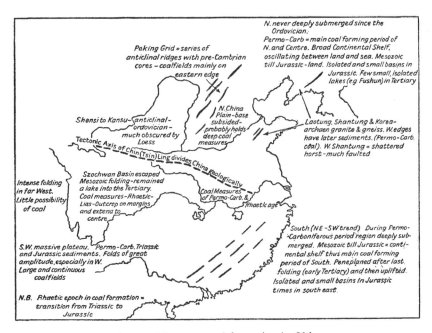

FIG. 37. Notes on coal formation in China

providing conditions suitable for coal formation. Thus it was that enormous coalfields were laid down over Kansu, Shensi, Shansi, Shantung, the North China Plain and Manchuria. From the end of the Permian, the north has remained above sea-level, except for small basins in which, during the Jurassic and Tertiary, thick coal beds were laid down as at Fushun and Fuhsin. These northern Permo-Carboniferous beds come to the surface in the following fields:

1. Coal is found in all three provinces of Manchuria. In Liaotung it is exposed along the western edges of the Irkhakhun Mountains and the Changpei Shan, which form the highland divide between Liaotung and Korea; also along the eastern edge of the Ta Chingan Mountains of Jehol and in many spots in between. There is very little anthracite in Manchuria, but there is a great wealth of bituminous coal, the reserves of which are estimated at about 17,000 million metric tons, more than half of which is in the two northern provinces of Heilungkiang (5,000 million tons) and Kirin (5,581 million tons). These two provinces have also considerable reserves of lignite, estimated at about 4,500 million metric tons. At Fushun and Fuhsin in Liaoning are the two phenomenal Tertiary seams, over 400 feet thick and overlain by oil shales. Neither in quality nor quantity do the Manchurian fields compare with those of Shensi and Shansi, yet before 1949 they were by far the most heavily worked of the Chinese fields. Until 1946 they were under Japanese control and were developed for local and Japanese needs. Production was stepped up from 13 million tons in 1936 to 30 million tons in 1944.

2. The east and west exposed fields of Liaotung are continued southward into China Proper, on the one hand, along the much faulted western edge of the Tai Shan of Shantung, and on the other, along the east-facing scarp of the Heng Shan and Tai Hang Shan in Hopei. In all probability these are the outcropping edges of a continuous bed deep under the continually sinking geosyncline of the North China Plain. The deepest recorded borings to date (2,840 ft.), however, have not penetrated below the alluvial deposits.[1]

The Shantung field has very little anthracite, but has an estimated reserve of 1,613 million metric tons of bituminous coal, which Richthofen reported as black, hard, burning with a clear flame and making excellent coke. This field was linked by rail with Tsingtao after the occupation of that port by Germany, and is reported to have produced more than 10 million tons in 1944.

[1] L. Carrington Goodrich, *China* (Los Angeles 1946), p. 42.

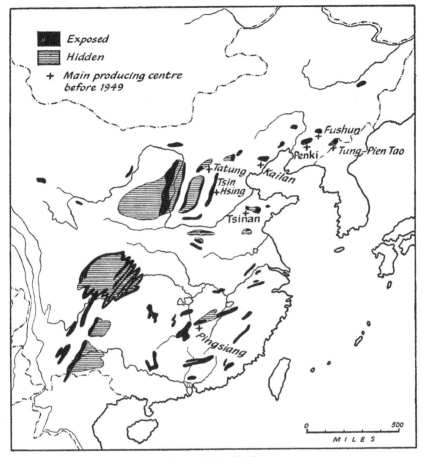

FIG. 38. Coalfields

On the Hopei-Honan border of the North China Plain, along the edge of the Shansi plateau, there are considerable reserves of both kinds of coal—5,430 million metric tons of anthracite and 5,400 million tons of bituminous coal. The main anthracite beds are in the south and extend into Honan across the Hwang-ho. The bituminous beds farther north provide good coking coal. This field was the first to be developed by modern techniques in China through the energy of Governor Li Hung-chang, one of the ill-fated progressive group. It was developed initially in response to the demand for coal for the newly inaugurated indigenous China Merchant Steam Navigation Company, for which coal from South Wales was being imported. Later it supplied the needs of, and was served

147

by, the Peking-Hankow Railway. The first shaft was sunk at Kailan in 1878 but it was not till 1886 that permission to build the railway was granted, one of many examples of the opposition which early industrialization had to meet. The railway runs parallel to the edge of the Shansi plateau into which branch lines penetrate to the many coal basins located there. Until 1949 almost the entire production from this field was carried out by two large concerns: the Kailan Mining Corporation and the Peking Syndicate.

3. Westward from Shansi into Shensi and Kansu is a vast, rich field, which, although largely obscured by Mesozoic Sandstones and loess deposits, is exposed on both sides of the Hwang-ho from Paotow to Sian. Here are China's greatest reserves, estimated at 37,443 million metric tons of anthracite, 160,466 million tons of bituminous coal and 3,727 million tons of lignite. The bituminous reserves are largely in the west and are of medium quality. The seams are generally horizontal and easily mined. The anthracite field is in the east. Here again there has been little folding, and the seams lie horizontally until the Taihang plateau edge is approached, when they dip steeply. The main seam varies in thickness from 10 to 40 ft., having an average thickness of 18 ft. and is generally of high quality, hard, pure and low in ash.   ·

This is a region important for its potentialities rather than its achievement. Although the quality of coal is good and there has been comparatively little folding in this area, so that the thick seams are largely undisturbed and easy to mine, yet production, up till 1949, has been small relative to the more accessible fields to the east. Capital expenditure on railways has been prohibitive and too precarious, so that this field has remained virtually undeveloped.

Land lying south of the Tsinling line, for the most part, was deeply submerged during Palaeozoic times when the coal measures of the north were being laid down. In Triassic times there was a general uplift throughout China. The north became dry land and the south a region of shallow seas and continental shelf. It was during the Rhaetic period, i.e. the period of transition from Triassic to Jurassic, that the main coal beds of the south were formed.

4. In the lower Yangtze basin and in the south-east the coal beds are widely scattered and isolated and are small as compared with those of the north. Central China has both Rhaetic and Permo-Carboniferous measures. The

coalfields of Kiangsu and Anhwei, although low in reserves, have been important in supplying much of Shanghai's industrial power. There is anthracite in the west of the field and bituminous coal in the east. Working is not easy as there has been considerable disturbance by igneous intrusion.

There is little coal in Hupeh. A small quantity of anthracite is mined at Shihhweiyao in connection with the big cement works.

The Ping Hsiang field in Hunan, which supplied good coking coal to the Hanyang Iron Works at Wuhan at the turn of the century, is the most important working in central and south China. These are Rhaetic measures. 'Thick seams, up to 36 ft., produce soft coals, mostly slack and high in ash but this soft, slack coal produces good coke, low in phosphorous and sulphur and eminently suitable for blast furnace practice.'[1] Total reserves of the Lower Yangtze, i.e. Hunan, Hupeh, Kiangsi, Anhwei and Kiangsu, are: anthracite 1,132 million metric tons; bituminous coal 1,773 million metric tons.

5. The Red Basin of Szechwan remained a lake into the Tertiary. Cretaceous red sandstone overlies the Rhaetic beds which underlie the whole basin and which outcrop around the edge, extensively along the crests of the anticlines in the east and south. Coal in the north is of better quality than in the south, but Szechwan coal generally is inferior to that of the rest of the country. There are some Permo-Carboniferous measures in the Ta Pa Shan, which form the north-eastern border of the province. There is little likelihood of coal being found in the intensely folded highlands to the west of the Red Basin. Szechwan's estimated reserves are: anthracite 293 million metric tons: bituminous coal 3,540 million metric tons.

6. Extensive Rhaetic coal deposits are found in the south-west plateau of Kweichow and Yunnan and in West Kwangsi, totalling 944 million metric tons of anthracite, 4,346 million tons of bituminous coal and 694 million tons of lignite. Because these fields are remote and lacking in modern communications, little use had been made of them before 1949.

It is generally true to say that the coal deposits of China decrease both in quality and quantity southward. There is also some truth in the generalization that methods of extraction are cruder and more primitive in the south. Far too many small undertakings lack the rudiments of safety—pits with little or no propping, the merest of ventilation and drainage, and a complete reliance on human power.

[1] W. Smith, *Coal and Iron in China* (Liverpool 1926).

Coal Reserves and Production[1]

Province	Reserves 1934–45 (figures in million metric tons)				Production 1944 (figures in thousand metric tons)
	Anthracite	Bituminous	Lignite	Total	
Heilungkiang		5,000	3,980	8,980	3,047
Kirin		5,581	478	6,059	6,117
Liaoning	36	2,606		2,642	10,940
Jehol		4,714		4,714	5,359
Chahar	17	487		504	9,300
	53	18,388	4,458	22,897	34,763
Shantung	26	1,613		1,639	10,300
Hopei	975	2,088	2	3,965	12,000
Honan	4,455	3,309		7,764	300
	5,456	7,008	2	13,368	22,600
Shansi	36,471	87,985	2,671	127,127	6,250
Shensi	750	71,200		71,950	650
Kansu	59	997		1,056	110
Ningsia	173	284		457	140
	37,943	160,466	2,671	200,590	7,150
Hupeh	45	309		354	40
Hunan	741	552		1,293	550
Anhwei	60	300		360	1,250
Kiangsi	271	420	9	700	120
Kiangsu	25	192		217	1,100
	1,142	1,773	9	2,924	3,060
Szechwan	293	3,540		3,833	2,700

[1] Compiled from N. Dickerman, *Mineral Resources of China.*

Yunnan	77	1,539	694	2,310	260
Kweichow	822	1,696		2,518	250
Kwangsi	45	1,111	1	1,157	200
	944	4,346	695	5,985	710
Chekiang	22	78		100	2
Fukien	147	6		153	30
Kwangtung	59	274		333	100
	228	358		586	132
Sinkiang		31,980		31,580	180

IRON

We have already seen that iron was beginning to oust bronze in the middle of the Chou dynasty (*circa* 700 BC). It was not until the end of that dynasty or the beginning of the Han that it had really established its ascendency and was being brought into ordinary everyday use.

All smelting was done by charcoal in innumerable small local furnaces, and since the cost of transport by coolie was prohibitive, production was simply for local needs of a domestic character, such as cooking *ko* (pans), scissors, agricultural implements, such as hoes, tips to ploughs, sickles, etc. The iron industry remained in this primitive state right into the twentieth century, early in which it was estimated that the Chinese consumption of iron per head per annum was a mere 3 lb. as compared with 285 lb. in Great Britain and 550 lb. in the USA.

Iron ore deposits are very widely distributed throughout the country, but all geological surveys up to 1949 are agreed that most fields contain ores of low iron content (30–40 per cent) and high impurity. Estimates of iron ore reserves have varied considerably. An early National Geological Survey estimated the reserve at 979 million metric tons, while the same body in 1945 placed the reserve at 2,184 million metric tons. A Japanese estimate for Manchuria alone, two years earlier, gave the figure of 3,803 million metric tons. Figures for reserves are misleading unless the various grades of ore included are clearly stated. Moreover, with changing knowledge in the use of ore and in technology, for example the processing and concentration of ores before blasting, it may be legitimate to include lower grade ores which formerly were rightly excluded.

FIG. 39. Iron ore resources

Be that as it may, expert opinion has been agreed that, while China has ample reserves of coal, her reserves of iron ore are such that she could not reasonably hope to develop industrially on the same massive scale as the USA or even Great Britain. Dr. J. S. Lee in his *Geology of China* says 'It is quite clear that China can never be an iron producing country of any importance'. George B. Cressey says 'It is clear that China cannot equal the industrial countries of the West in the production of iron'.[1] R. H. Tawney writing in 1932 says 'As far as the visible future is concerned the limiting factor is not coal, but iron . . . The growth of an iron and steel

[1] G. B. Cressey, *Land of 500 Million* (New York 1955), p. 140.

industry of considerable dimensions is hardly possible in China'.[1] He goes on to advise that Chinese industrial developments go along the lines the French followed about 1910–14, i.e. developing light rather than heavy mass production and so using those skills, dexterity, craftsmanship and taste most characteristic to her genius. It is important to bear in mind this widely held pre-1949 assessment and judgement in view of subsequent policy and development.

The first large scale modern iron and steel smelting plant in China was built at Hanyang at the confluence of the Yangtze and the Han Rivers. Its location there was fortuitous. Chang Chih-tung, Viceroy of Kwangtung (1884–9), was another of the ill-fated progressive group which came to grief at the hands of the Empress Dowager at the end of the century. He had ordered a complete iron and steel smelting plant from England, which he intended to erect at Canton. However, before it arrived he was appointed Viceroy of Lianghu (Hupeh and Hunan) and so he promptly re-routed his machinery to Hanyang, which lies on the opposite side of the Yangtze to Wuchang, his official seat. Here he constructed the Han Yeh Ping Iron Works, the primary purpose of which was to supply rails for the construction of the Peking-Hankow Railway. This siting of the plant was fortunate. About seventy miles downstream from Wuhan is the iron ore of Tayeh, a fine 60 per cent haematite along a limestone contact with diorite. Great difficulty was encountered in securing supplies of coke for the blast furnaces. At first coke was shipped from Europe, and later from Kaiping in north China, but eventually good coking coal at Ping Hsiang, Hunan, was discovered. This could be shipped through the Tungting Lake to Hanyang. Even so, coke formed 50 per cent of the cost of production at the iron works.

The Han Yeh Ping Works had a very chequered career. At first it enjoyed some government protection and tax exemption, but later it suffered from continual internal political unrest and instability. Production was never very great. In the 1920's Japan gained financial control of the works. Japanese interest lay rather in securing control of the iron ore of Tayeh for shipment to Japan than in developing steel production at Hanyang. The works were allowed—even induced—to fall into decay. Their final demise came with the bombing during the Sino-Japanese war, and no attempt was made at their resuscitation before 1949.

This development of the iron and steel industry in Central China was very small as compared with that in Manchuria by the Japanese. With the defeat of the Russians in the Russo-Japanese war (1904–5), Japan took

[1] R. H. Tawney, *Land and Labour in China* (London 1932).

Iron Ore Reserves and Production

	Reserves 1934–45 (figures in thousand metric tons)	Production 1942 (figures in metric tons)
Heilungkiang	500	—
Kirin	15,700	—
Liaoning	1,390,050	4,413,306
Jehol	11,340	—
Chahar	93,645	923,376
	1,511,235	5,336,682
Shantung	15,340	32,056
Hopei	42,179	—
Honan	21,536	25,000
	79,055	57,056
Shansi	34,230	82,000
Shensi	10,847	1,800
Kansu	2,496	2,400
Ningsia	7,579	—
	55,152	86,200
Hupeh	143,174	1,454,828
Hunan	31,753	16,000
Anhwei	19,204	1,481,000
Kiangsi	15,466	1,300
Kiangsu	5,700	—
	215,297	3,059,328
Szechwan	22,023	128,020
Yunnan	12,156	25,129
Kweichow	40,553	13,000
Kwangsi	2,067	8,000
	54,776	46,129

Chekiang	3,224	300
Fukien	92,562	6,100
Kwangtung	52,155	1,013,902
	147,941	1,020,302
Sinkiang	48,737	—

over, under protest from the Chinese, the concessions in south Manchuria, which the Russians had obtained from China as a result of the Boxer Uprising. These concessions included control of the railway from Port Arthur to Changchun, which Japan proceeded to develop into a network as a basis for industry. The main expansion, however, came after Japan established the satellite state of Manchukuo in 1931.[1] The great Showa Iron and Steel Works at Anshan were expanded: the nine blast furnaces had a maximum capacity of 2,500,250 metric tons of pig-iron and 837,000 metric tons of steel per annum, although actual production never reached this figure. Japan inaugurated a Manchurian Five Year Development Plan in 1936, but iron ore proved a stumbling-block to its achievement by failing to keep pace with the expansion of iron and steel plant.[2]

There are three main iron ore fields in Manchuria. Two, the Anshan and the Penhsihu, lie in the heart of the Liaoning plain. The third at Tungpientao is near the Korean border by the Yalu River. The Anshan field is productively the most important, although its ores are mainly low grade (35 per cent) magnetite and haematite. There is a small amount of high grade (60 per cent). Output in 1943 was 3,125,000 metric tons or 59 per cent of the Manchurian total. The Penhsihu field in the same year produced 1,125,000 tons or 21 per cent of the total. Both fields have ore, coal and limestone in close proximity. The eastern Tungpientao field has both high and low grade haematite ores, again near coal. Output in 1943 was 849,933 metric tons of ore. The Japanese were developing this field as a minor iron and steel centre in 1945 when the war came to an end.

With the defeat of Japan heavy industry in Manchuria came to a standstill. Soviet authorities removed a great deal of the heavy plant to Russia. Civil war tore the country until the Communist forces gained control in 1947 and reconstruction commenced.

[1] T. Read, 'Economic-Geographic Aspects of China's Industy', *The Geographical Review*, Vol. 33, 1943.

[2] A. Rogers, 'Manchurian Iron and Steel Industry', *The Geographical Review*, 1948, p. 41.

OIL

Oil surveying and prospecting before 1949 gave little promise of large reserves being found anywhere within China's borders. Possible productive fields were known in Kansu, in Sinkiang and in Dzungaria. A small amount of oil had been actually produced at Yumen, Kansu and at Tihwa, Dzungaria, but not in sufficient quantity to be significant commercially or to warrant the necessary modern means of transport being constructed.

Of much greater importance and promise was the oil produced from the shales overlying the open-cut coal measures at Fushun. Some 200 ft. of oil shale and 200 ft. of green shale constitute the overburden of the coal which must be removed before the coal can be extracted. The Japanese built an oil refinery immediately beside the coal cut; in fact so close that workings are now endangering the whole plant.

NON-FERROUS MINERALS[1]

China is fortunate in possessing some of the world's more important non-ferrous minerals in relative abundance. These minerals are concentrated mainly in the south-west corner of China; consequently they are often remote and development has been handicapped by poor communications.

1. *Antimony.* Although in small supply, antimony is an important alloy in a number of white metals, e.g. pewter, bearing metal, Babbitt's metal, and is used in casting type and battery grids. In 1933, when world production stood at about 22,000 metric tons, China produced about 68 per cent; 17 per cent came from Bolivia and Mexico and the remainder from the USA, Czechoslovakia and Yugoslavia. Chinese supplies were cut off during the Sino-Japanese war with a result that American production was stepped up. By 1949, China's place in the world market had practically disappeared. World production then was 34,000 tons of which the USA produced 12,500 tons, Bolivia 10,900 tons and Mexico 7,000. China's supplies in the future are all likely to be held for home use. Formerly they were exported via Hankow, where they were processed, to Shanghai.

China's main reserves and production are at Hsikwangshan, west of Changsha, Hunan. There are smaller reserves in Kweichow, Kwangtung, Kwangsi and Yunnan.

2. *Bauxite.* Bauxite, the basic ore for aluminium, is to be found in plenty in Yunnan, Shantung, Liaonong and Kansu. Cheap and plentiful electric

[1] Wang Kung-ping, 'Mineral Resources of China with Special Reference to Non-ferrous Metals', *The Geographical Review*, Vol. 34, 1944.

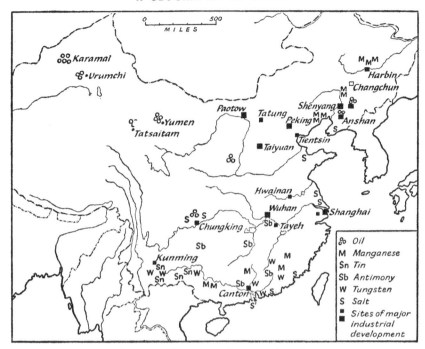

FIG. 40. Oil and non-ferrous minerals

power is necessary for the production of aluminium, and up to 1949 this was available only on the Liaoning coalfields at Penchi and Anshan and also at Tainan, Taiwan. Exploitation of bauxite resources is likely only after the development of large scale hydro-electric works.

3. *Copper.* China's reserves of copper are not large but they are widespread. There are a few first-class deposits. In terms of world production, China's contribution of a few hundred tons per annum during this century is negligible. Apparently this has been about the usual annual production through the centuries, for the average output between AD 847 and 859 was 290 tons. It is remarkable that a country which has used a copper coinage for more than 2,000 years should produce so little.

The most promising fields are in West Szechwan where there is an estimated reserve of 10 million metric tons of sulphide ore. Tungchwan in north-east Yunnan is reported to have deposits of 3–10 per cent ore. South-east Sikang has estimated reserves of 4½ million tons. Tienpaoshan, Kirin, has ores of 1·7 per cent copper, 6 per cent zinc, 5 per cent lead and some silver.

4. *Gold.* Gold is found in north Manchuria (Heilungkiang and Kirin) and in west Sinkiang. In both areas it is mined as placer deposits mainly in rivers and streams. It is a poor man's occupation and its production, never great, is decreasing.

5. *Mercury.* The production of mercury is one of China's oldest industries, having been practised for more than 2,000 years. Cinnabar is found in west Hunan, Kweichow and south-east Szechwan. Production has steadily declined. In the nineteenth century over 1,000 metric tons were produced annually. This figure fell to between 130 and 470 metric tons between 1914 and 1925.

6. *Salt.* The foreigner coming to China earlier in this century would be struck by the care with which salt was treated in the countryside. Salt was a government monopoly, heavily taxed and therefore expensive. It was obtained mainly through evaporation in shallow pans, works being situated along the whole length of coast. The saline basins of the north-west also produced their quota. In Szechwan there are many salt wells which have operated for centuries.

7. *Tin.* The South-west Plateau of Yunnan shares with Malaya and Burma the world's richest tin deposits. Estimates of reserves vary very widely between half a million and two million tons. The main producing centre is Kochiu, South Yunnan, which is accredited with 80–90 per cent of the total output. Production has fluctuated considerably during the century, owing partly to unsteady world prices and partly to unrest within China itself.

	metric tons
1920	10,000
1930	6,500
1938	15,540
1941	12,000
1950	7,000

Tin is important because of the large part it plays in many alloys:

Alloy	% Tin	% Copper	% Lead	% Others
Bell metal	15–25	85–75		
Coinage	4	95		1 Zinc
Babbitt's metal	90	3		7 Antimony
Bearing metal	75	12·5		12·5 Antimony
Pewter	82			18 Antimony
Type	26	1	58	15 Antimony

PLATE X: The high mountain valleys of the west and north-west are the grazing
lands for livestock, especially sheep and goats.

PLATE XI: Artificial ponds are stocked with freshwater fish. Nearly every village has its own pond and, particularly in the south, the ponds are on a relatively large commercial scale.

PLATE XII: Anshan Integrated Steel Works. Over 4 million tons of steel a year are produced by the Anshan Steel Works, which is the greatest integrated centre in the country. Other big centres are now at Paotow and Hankow.

8. *Tungsten.* Wolfram or wolframite, the ore of the metal tungsten, is the most important of China's non-ferrous minerals. The main ore field is in the south-centre (South Kiangsi, South Hunan and North Kwangtung), which has reserves of about 1 million metric tons of high grade ore. A second field is a belt running the length of the coast from Fukien to Hainan.

Since 1915, when the deposits of South Kiangsi were discovered, China has been the world's chief tungsten producer. As with tin, production has fluctuated and for the same reasons, the annual output ranging from 5,000 to 18,000 metric tons per annum.

Tungsten is used as a constituent for high-speed steels, metal cutting tools and in the filament of electric light bulbs.

HYDRO-ELECTRIC POWER

While China has, in her great river systems, enormous potential water power, Manchuria was the only region of China, before 1949, where any hydro-electric power had been developed on a large scale. After their seizure of Manchukuo in 1931, the Japanese had ambitious plans, which they had partly succeeded in implementing before their defeat in 1945. Large dams were built in the north on the upper Sungari and on its right bank tributary, the Mu-tan Kiang, and on the left bank tributary, the Nonni River, at Tsitsihar. In the east, on the Yalu River at Supung, another large dam was constructed. In 1945 the annual generation of electric power was about 3 million kilowatts. This figure included nearly 300,000 kW. of thermal electric power, generated at the open coal cut of Fushun.

ECONOMIC DEVELOPMENT AFTER 1949

It is difficult, if not impossible, to comprehend the change that has taken place in China during the twentieth century. Many of the changes that have taken shape in the last decade or so have been in preparation for a century or more, but few people, if any, can have anticipated the suddenness of their fruition or their revolutionary immensity.

Probably more far-reaching and fundamental even than China's ultimate embracing of Communism has been the decay and final collapse of its ancient civilization based on Confucian philosophy. Although that decay had started before Western civilization had begun to force itself upon the Celestial Empire, it was that Western intrusion which greatly accelerated the process. Western science and the power—military and economic—which it gave, and Western ideas of individual freedom, together helped to bring tottering Confucian conservatism to the ground. All the elaborate protocol between high and low; all the careful code of behaviour within the family, its emphasis on filial piety, its rules of marriage and its strict code of morality—all these were fading fast in the chaos of the early decades of the twentieth century. But Western civilization, both by the moral bankruptcy which it had revealed by waging two world wars and in the imperialism by which it had imposed its will on an impotent China for a century, was unable to fill the spiritual vacuum it had helped to create. The self-sacrificing and altruistic work and the material wealth in hospitals, schools, etc., poured into the country by Christian missions for a century, were so identified in young China's eyes with an overbearing West, which had brought constant humiliation, that they brought no commendation. Western civilization came to be regarded as synonymous with imperialism and all that was evil, and something to be cast off.

So it was that China's leaders and reformers looked elsewhere for a pattern on which to mould their future. Sun Yat Sen, in spite of his Western education and the fact that he owed his life to the freedoms the West maintained, had nothing but condemnation for the West's treatment of his country. He turned to young Soviet Russia, just rising from bondage of the capitalist West as he saw it, for friendship and inspiration. The Kuomingtang (Nationalist Party), which was his creation and which was in alliance with Russian Communism in the victorious march northward from Canton under Chiang Kai-shek in 1926–7, overthrew the war lords. However, the Kuomintang was soon at loggerheads with a nascent Chinese Communist party, under Mao Tse-tung's leadership. There then followed a long and bitter struggle between the two, in which there was a nominal truce in the later stages of the Sino-Japanese war, when they united against a common foe. After 1946 the struggle was renewed. The Kuomingtang became steadily more decadent, more corrupt, more inefficient and more reliant on foreign funds from UNRRA and particularly those from the USA. So it became identified in Chinese eyes with the Western powers, as its 'running dog', and when it fell before the onslaught of the Communist forces from the north in 1949 with an ease that surprised even the victors, it was almost universally discredited and unsung.

At the beginning of 1949 China was nominally, at least, under a government of a Western pattern, and its economic life, however chaotic and disorganized, was based on the private ownership of capital and on individual enterprise. By the end of that same year its rulers were the Chinese Communist Party, which had as its avowed ultimate objective the establishment of a communist society and as its immediate objective, in the words of

Chou En-lai, 'the release of the productive powers of the country from the oppression of imperialism, feudalism and bureaucratic capitalism', which they forthwith began to implement at a speed and energy which has amazed all observers, although it may not have met with their approval.

The advent of the new régime was accompanied by a surge of nationalism and patriotism which particularly captured the young, many of whom, in their enthusiasm threw over loyalty to family in favour of loyalty to state. Communism, too, brought with it a puritanism and new social standards, the quick acceptance of which was startling. Betting and gambling, so universal in China, disappeared overnight; prostitution, begging, stealing, suddenly dwindled as being contrary to 'the good of the people'.

Such were some of the fundamental changes which rocked the country in the early years after the revolution, and such is the setting and psychological climate in which the economic development of the last decade must be seen.

Rehabilitation (1949–1952)

When, at the end of 1949, the Chinese Communist Party (CCP) unexpectedly found themselves masters of the whole country, they inherited difficulties and problems on all sides of life. After years of war and civil war the country was in a sorry state. There was much banditry in the countryside and the first job was to restore law and order. The well disciplined communist forces, by their considerate behaviour, which contrasted sharply with that of the Kuomintang troops, did much to reassure the common folk and to recommend the new régime. For the first time since 1911 the country was brought under one rule.

War and civil war had brought production to a very low ebb, so that in 1949 the country faced imminent danger of famine. The whole economy was disrupted and practically at a standstill. China's monetary system was virtually non-existent. Between 1946 and 1949 there had been rapid and disastrous inflation, which the Kuomintang had tried to meet by two currency 'reforms'. The second of these reforms took place in September 1948, when the Government called in, under dire threats, all silver dollars, the only acceptable money then circulating, and issued paper 'Gold Yuan' at the rate of two G.Y. for one silver dollar in exchange. In nine months these Gold Yuan had fallen to less than one millionth of their value. This inflation was one of the major causes of the fall of the Kuomintang, and was one of the biggest immediate problems the CCP had to solve. When its new *Jen Min Piao* (People's Notes) were issued, another inflationary period was generally anticipated. However, by vigorous government control, prices and wages were tied to an index, fixed by the price of five commodities (rice, oil, coal, flour and cotton cloth). Consumption was reduced by an austerity campaign, by high taxation, by the issue of bonds, the purchase of which was virtually compulsory and also by a careful control of the amount of money in circulation. Government or state trading companies were formed and used to stimulate production, while private firms were forced to unload their stocks and so help to prevent 'too much money chasing too few goods'. For two to three years considerable hardship was experienced, but the greater evil of inflation was avoided, and in spite of the severe strain of the Korean War, the currency was stabilized.

The new government also turned its attention quickly to communications. By 1949 both rail and water transport had deteriorated to a shocking extent. The permanent way was such that a rail journey was a nerve-racking event. As one's train crawled over ominously creaking temporary structures that passed for bridges, one peered at half-sunken remains of former, less fortunate, attempts below. Passenger and goods transport by river steamer was spasmodic, unreliable and filthy. This breakdown of communications, coupled with the disruption of trade caused by inflation, meant that there was little flow of commerce between the various parts of the country and a greater fragmentation of the economy than usual. Realizing that, if the existing meagre resources of the country were in any degree to meet the pressing needs, there must be much better facilities for exchange between the various parts, the government set about the rehabilitation of the country's railways with great vigour. Within two years tracks were relaid and bridges rebuilt so that journeys which had been a nightmare, were both safe and comfortable. Foodstuffs and raw materials for industry were beginning to flow.

By 1952, production in the main foodstuffs and raw materials had just about caught up with the peak levels of production before 1949. During these years of recovery, plans were prepared for the inauguration of the First Five Year Plan. However, before this plan is examined, we must turn our attention to developments in agriculture.

Land Reform

We have seen the abject state of both peasant and agriculture before 1949, the fragmentation of the land, his scattered, small plots and holdings, the heavy rents, the increasing absentee landlordism, his traditional and often inefficient methods and technology, his primitive tools, poor seed selection, his restricted and controlled markets, his lack of credit and indebtedness to money-lenders, the oppressive taxation by war-lord or legitimate government. Add to all this the natural hazards of flood and drought, which, without stable government to control them, occurred with depressing and disastrous regularity, and the peasants' cup of poverty and misery was full. These are the conditions which had arisen many times before in Chinese history, and these are the conditions which set the stage for either peasant revolt, or sweeping reforms, or both.

It was into a countryside ripe for revolt that the newly-formed army of the young Kuomintang, under Chiang Kai-shek, marched from Canton in 1926. The peasantry received them with open arms and acclaim, for they were preceded by propaganda promising land reform, Sun Yat Sen's 'land to the tillers' and many far-reaching agricultural changes. The war lords' armies melted before it. In 1930, a comprehensive programme was drawn up, embodying some measure of land reform, fixing rents and assuring security of tenure, co-operatives with credit and marketing facilities, land reclamation, water conservancy and irrigation. However, the KMT failed signally to implement the schemes, partly because of the Japanese invasion, but mainly through power-seeking and corruption within their own ranks.

When, in 1949, the Chinese Communist Party (CCP) forces advanced southward against the KMT, they were preceded by propaganda, which, on the agricultural side, was almost identical with that which the KMT had used in 1926. Again the peasantry welcomed the invaders and again opposition crumbled, but this time there was a difference. The CCP in the long and bitter years between 1927 and 1949 had had intimate experience among rural people—an experience which has had a profound influence on their ideology and their communism. Unlike the KMT they wasted no time in giving effect to their land reform promises, which were in fact an integral part of their long-term policy. On the political side it gave effect to, and was a stage in, the class struggle, which the CCP initiated, for it meant the elimination of the scholar-landlord-administrator class. Doubtless the new rulers remembered their history—that in all the many peasant revolts which ushered in new régimes or new dynasties, the ruling scholar-landlord class was unseated only temporarily and quickly climbed back into power—and they saw to it that this history was not repeated. The rural population was divided into five classes—landlords, rich peasants, middle peasants, poor peasants and workers or the landless—and the order was issued for the re-distribution of the land. Holdings were taken from the landlords and rich peasants and re-apportioned so that the members of families in any given district, held more or less the same amount of land per head. The holding per head in the wheat-growing north was considerably greater than in the rice-growing south. This re-distribution was not carried through without a good deal of disorganization, bitterness and violence, and was not completed throughout the whole country until 1953, although it was finished early in 1952 in the north.

Chou En-lai had spoken of the revolution as the 'release of the productive forces of the country from imperialism, feudalism and bureaucratic capitalism'. Land reform was intended as a first step in the release from feudalism but only as a first step. The CCP had made no secret from the outset that their ultimate objective was the establishment of first a socialist and then a communist society. There would be no individual ownership of the means of production, including land, although this was probably not appreciated by the peasant, who had just received the title deeds to his land. It was probably the government's intention not to interfere with peasant ownership for some time, but events dictated otherwise. The overriding and compelling objective in all immediate CCP policy—agricultural, industrial and even political—has been, and still is production, production and ever more production. Land reform and re-distribution clearly meant greater fragmentation of the land and with it, often, ownership and working in the hands of less able farmers. The immediate result was less production rather than more, even allowing for the greater enthusiasm and energy that came from new ownership. Therefore, in 1954, the millions of new proprietors were organized into Temporary (and later Permanent) Mutual Aid Teams, that is to say, groups of five to ten families were banded together in a team to work each other's land. This was no great innovation since it had been practiced long before, but never on this scale or with this organization. The peasant retained the ownership of his land, which was worked by the team as he directed, his labour and tools being pooled for the season and remunerated on a point system.

In the following year, 1955, the first steps were taken in the next stage of agricultural organization: the formation of the Elementary (and later Advanced) Agricultural Producers' Co-operatives. Considerable opposition was met with at this stage, for it involved the peasant contributing his recently acquired land and his stock, tools, carts and draught animals to the common pool to be worked as a whole under an elected management committee. He was rewarded for his labour on a points system from the produce of the co-operative and he also received some return in respect of his capital contribution. Participation in the co-operative venture was voluntary and secession permitted, but a good deal of social pressure was brought to bear and refusal to join was not always easy. Management was intended to be democratic, but in fact the ubiquitous Party cadre often wielded virtually dictatorial powers. This bureaucratic control is a recurring source of complaint, and calls from above for its correction have constantly been issued. That the right of secession was not entirely nominal at this stage is evidenced by the withdrawal from their co-operatives of about 100,000 upper middle class peasants in Kwangtung in protest against a fall in income below the level they had enjoyed before joining.[1] Many co-operatives were closed at this time but were quickly re-started, until by the end of 1957 practically the entire agricultural population were members.

The size of the co-operative and the number of members or families included varied considerably in different parts of the country. The average was about 160 households or some 600 persons, who were organized into Production Brigades and Production teams. This number was criticized as being too large, lending itself to bureaucracy and impersonal relationships detrimental to production, and in fact a decision had just been taken at high level to reduce the size of the co-operative when the whole agricultural organization underwent yet another radical change in the formation of the People's Communes.

In April 1958, between twenty and thirty higher co-operatives, comprising some 20,000 members in the province of Honan combined to form themselves into one unit, which they called a Commune. The aim in so doing was probably economic, to increase production, rather than political, although doubtless the keen Party members amongst them saw in it a further step in ideological attainment. By so combining, it was held, the commune could undertake all the local government duties of the *hsiang*[2], which it would replace, and also, because of its increased size, the commune would be better able than the individual co-operative efficiently to organize all the agricultural and industrial activities of the area. It could cope with water conservancy and irrigation on a much more comprehensive scale than the smaller co-operatives. It would be responsible for all educational and cultural work, and also for the police and local militia. In fact, the commune would be the authority controlling and directing the life of the district in all its facets. In August 1958, Chairman Mao Tse-tung visited and inspected this Honan experiment, found it good and gave it

[1] T. J. Hughes & D. E. T. Luard, *The Economic Development of Communist China, 1949–1958* (Oxford 1959), p. 149.
[2] *Hsiang*—a district of not more than 50,000 inhabitants.

his blessing. This was the signal for communes to be started immediately throughout the land. In September 1958, there were about 700,000 Higher Agricultural Producers' Co-operatives in existence. Within a year these had all been merged into 26,000 People's Communes.

The organization and running of communes differs considerably from place to place, but there is a general pattern which is common to them all. The commune is the supreme body, whose function is to arouse and inspire the life of the community, to 'organize their activities with high militancy and to improve their productive efficiency and labour discipline'. On the economic side it is responsible for the formulation of production plans, which are submitted to the State Planning Organization for approval. It is also responsible for the sale of products, the purchase of machinery and for capital accumulation and construction. The commune owns only a small part of the means of production and is concerned mainly with the direct operation of small industrial establishments such as native chemical fertilizer plant and farm implement repairs plant. These industrial establishments are collectively owned by all the members of the commune. The commune council is made up of representatives elected by the production brigades.

The production brigades directly organize all production and all collective welfare undertakings. They consist usually of about 100 families or several hundred labourers and control about 1,000 *mow* or 170 acres. The brigades are independent operation and accounting units and own most of the land, draught animals and farm implements. It is the brigade's job to 'fix four things' for the Production Teams below them. They allocate the land to be worked, they fix the necessary labour force and they apportion both the draught animals and the farm tools.

The production teams, usually 20–30 families then agree with their production brigade, on the basis of these allocations, to guarantee a certain output or quota and also to guarantee to do so at a certain production cost. This guarantee or agreement is of great importance, for on it depends, to a very large extent, the remuneration that the team will receive. 'A production team's income is based on the guaranteed output. The brigade distributes "consumer funds" to all members, through production teams, after deduction of tax, production and administration expenses, reserve and welfare fund. The amount of consumer funds is not fixed, nor are issues generally made each month. The amount of consumer funds provided by the crops each year is distributed according to the work points earned by each person, and is issued to commune members several times a year. A final settlement is made after the autumn harvest or at the end of the year. This system is essentially the same as the wage system, distribution according to work, "more pay for more work".[1]

The production team, by over-production of its guaranteed output, can earn bonuses. After fulfilling their quota the members of the team can further add to their income by various side occupations, such as fruit growing, breeding pigs and poultry, bee-keeping on their own small-holdings and this they are urged to do. In the early stages of the commune this was not so. 'All means of production of co-operatives are turned over to common ownership of the commune, including all the small plots of private holdings, privately owned house sites and other means of production, such as livestock, tree holdings, etc. They are allowed to keep a small number of domestic animals and fowl as private property.'[2] This drastic measure was modified quite quickly. At first, also, there was a great deal of officious direction of the production team from above, with little regard for local knowledge or local conditions, leading to discontent, non-co-operation and a fall in production, as the following extracts indicate:

'What are the characteristics of a particular plot? What is the quality of the soil? What are the crops suited to the land? How should sowing be arranged? Over these questions, only those team members and cadres who are most familiar with local conditions have the right to speak. Under no circumstances may other people, regardless of the concrete circumstances, require production teams to plant crops that are not suited to the concrete conditions, though such people may have good intentions. If they do so, even high-yield crops will frequently be turned into low yield ones.

'If such powers of production administration are not delegated to the production teams, if their rights are not respected, if they are not allowed to make their own decisions, and if everything is managed from the production brigade level, it will be impossible to call forth the initiative and enthusiasm of the production teams and impossible to arrange production according to the land, the seasons, crops and manpower. It will do great harm to agricultural production.'[3]

Much use has been made of the phrase 'walking on two legs', the meaning of which is that, in addition to the development of large scale, integrated state industry, every effort should be made to make use of local rural skills and to develop widely dispersed small scale industry in the country. It was this aspect which was strongly emphasized by the early promoters of the commune. The commune, being a much larger unit, it was urged, would

---

[1] Lo Keng-mo, 'The Character of Rural People's Communes at the Present Stage', *Kung-jen Jih-pao* (Peking, 21st July 1961).

[2] *Peking Review*, No. 30 (Peking, 23rd September 1958).

[3] Shu Tai-hsin, 'The Function and Power of Production Teams in Production Administration', *Kung-jen Jih-pao* (Peking, July 1961).

be better able to promote and handle rural industry. This side of the commune badly misfired, largely on account of Chinese methods of industrial planning, which have relied, as we shall see later, so much on 'target' planning. The launching of the commune synchronized with the 'Great Leap Forward' of 1958, when high targets of iron and steel production for the country were fixed. The campaign was taken up with great enthusiasm. All over the country thousands of little home-made blast furnaces were built in backyards and courtyards, local iron ore was collected and pig iron of unpredictable quality produced in large quantities. This ill-conceived campaign diverted attention away from the intended local industry and more important, took far too many productive brigades and production teams from their primary function, the gathering of a bumper harvest, much of which was never garnered in consequence. The effects of this were severely felt in the disastrous harvests of 1959 and 1960. Since then corrective efforts have been made and rural commune industries are directed into more fitting and legitimate channels. Priority is given to the manufacture of agricultural machinery and implements, fertilizers, insecticides, processing animal fodder, i.e. trades that directly serve agricultural production. Emphasis is also placed on trades which are essential to members' livelihood, such as food processing, tailoring and shoe making, and also on the manufacture of traditional handicraft articles and export goods.[1]

Since the formation of the communes a great deal of larger scale water conservancy and irrigation work has been carried through, work which was beyond the competency of the smaller co-operative unit. Large earth retention dams have been made for more continuous and reliable irrigation of the fields. These have been constructed by local labour and with a minimum of capital expenditure. Communes have been able to cope more adequately with seed selection and pest control. Also they are more firmly placed in regard to marketing and credit facilities. Although there is some mechanization, this has not yet progressed very far and the main reliance is still placed on human labour, working with hand tools, although these are more plentiful and better made.

Widely dispersed and scattered amongst the communes are some 2,500 large state farms, having an average area per farm of about 2,000 hectares. These farms have been developed mainly on reclaimed marginal or waste land. Notable examples are those farms in Honan on the lands desolated by the breach of the Hwang Ho in 1938; the reclaimed floodlands around

---

[1] Tso Ch'un-t'ai, 'Commune Industry operated by Rural People's Communes' *Kung-jen Jih-pao* (Peking, 27th July 1961).

the Tung-ting Lake in Hunan; and the huge 'friendship' farm of 40,000 hectares in Heilungkiang on the Sungari River. The state farms are manned mainly by demobilized service men, numbering in all about 2,800,000.

By careful irrigation, soil improvement by application of artificial fertilizers and by seed selection, the production on these marginal lands is about that of the national average. Mechanization, even on these large farms, has not yet gone very far—about 40 per cent of all the cultivation —although it is much higher than that of the average commune.

The state farms are intended as pioneering models for the surrounding communes, teaching new methods and acting as pace-makers.

One big objective in the formation of both rural and urban communes has been the mobilization of housewives for productive work by instituting collective feeding. Communal kitchens form part of the collective welfare services of the commune, which also include crèches, medical care and welfare for the old and infirm. But the kitchens in many instances have proved unacceptable and have been discarded.

There is no doubt that all the above factors together make the commune a more efficient production unit than the co-operative. Its problems seem to lie rather in the field of human relationships than in economic efficiency. A United Nations ECAFE survey in March 1959 reported that China's people's communes 'appear to possess marked advantages from the point of view of technical organization of production . . . The question, which only the future can answer, is whether they provide adequate human incentives'.

Planning

In considering the economic development of China after 1949, it is essential to keep in mind the basic aims guiding all Chinese communist activity. The CCP objective is the building of a modern industrial state from a backward agricultural one and doing it quickly in the space of one generation. It aims at changing a capitalist economy into a socialist-communist economy and an individual peasant economy into a co-operative-communist economy in the same space of time.

We have seen the steps that have been taken in the agricultural sphere, where at least the structure of a communist economy has been achieved in the space of only five years. It now remains for us to see how far industrialization has been implemented.

The period 1950–52 was one of rehabilitation and a breathing space, during which plans were formulated for the big changes to come. In these years production, in so far as it was planned at all, was on the basis of annual targets for individual industries. Private enterprise continued to function alongside rapidly growing state concerns. It is estimated that in 1952 practically all agricultural output and at least one third of modern industrial output was in private hands. Production had caught up with the pre-1949 peak.

In 1952 the Committee of Finance and Economic Affairs formed the State Planning Commission under the chairmanship of Li Fu-chun. In 1953 this Commission launched the First Five Year Plan amid a host of obstacles and difficulties. Within the ranks of the CCP there could be but few who had experience of industrial planning and old capitalist managerial and executive business skill was not acceptable. This experience had to be bought the hard way.

Planning was mainly on the basis of annual targets. Although there certainly must have been a broad blue-print of the whole plan, it was never published. The trouble with target planning is that it too often lacks co-ordination and lends itself to schemes that are too rapid and too ambitious. The result is frequent bottle-necks, over-production in one field and under-production in another. These are faults common to all economic systems, but target planning tends to accentuate them and has been in part responsible for uneven progress.

Planning in China has been bedevilled by lack of accurate statistical information. By the very nature of the case there was little or no reliable data on which to work. An attempt was made in 1952 by the newly-formed State Statistical Bureau to take an accurate census so that at least numbers were known. In addition, much patient statistical work was done in the economic sphere by the Bureau, which was building a reputation for reliability and honesty, but in 1958 it suffered a severe set-back at the hands of over-enthusiastic communist cadres. In that year which claimed such phenomenal advances in production, it was later revealed that returns in both the agricultural and industrial fields were gross over-statements. Whether through desire to stand high in Party esteem by appearing to have achieved more than their assignment, or through fear of reprimand and criticism if they failed to meet the assigned targets, returns from co-operatives, factories and mines were inflated, giving an altogether false picture of the country's productive power. This falsification, which according to one local estimate amounted to about 30 per cent, has contributed not a little to the serious industrial and agricultural problems since 1958 and has rendered all subsequent figures suspect.[1]

[1] Choh-Ming Li, 'Statistics and Planning at the Hsien level in Communist China', *China Quarterly*, No. 9, 1962; *The Statistical System of Communist China*, University of California Press 1962.

A third obstacle to rapid industrialization was the lack of a trained labour force. In 1917 Russia was more advanced industrially and thus had more to build on than had China in this respect in 1953. China has had a great deal of help from the USSR in the form of a loan of Russian expert technicians and of training for workers, who have been very quick to learn and adapt themselves to the new techniques.

Poverty of communications has largely vitiated the country's wealth of raw materials in the past, which explains the emphasis that has now been placed on rail, road and water development.

Lack of capital can be met in two ways: by loans and saving. China could look only to USSR for loans.[1] From that source she received about £100 million in 1950 to be spent on Soviet manufactures and a further £400 million in 1954 to cover the cost of 156 enterprises, including many complete plants, imported from Russia. Considerable though this is, it is a very small proportion of China's capital needs if she is to achieve rapid industrialization. She has therefore had to look to her own savings to provide the capital. Since China is an agricultural country, it has had to be from her agricultural produce that that saving has come. It is difficult to over-emphasize the role of agriculture in China's industrialization. Mao Tse-tung, from the earliest days of the CCP, has had no doubt of it. In a little quoted utterance, he stated that if he were allotting points, he would give seven to the peasant and three to the urban worker. This assessment of agriculture has had a profound influence on the nature of the Chinese revolution and is the main reason for the difference between Chinese and USSR practice.

It is from agricultural produce that the food of industrial workers must come as they build up the capital goods, the mines, blast furnaces, steel mills and factories, while others build railways and roads, construct dams and dykes for water conservancy. It is from agricultural produce that the host of new teachers and students must draw their sustenance while learning the new techniques, and it is from agriculture that the wherewithal must come with which to pay for needed materials from abroad.

In a country where so much of the population has been living so near to the subsistence level, there is a real danger that planners may attempt to save too much. They may try to build up capital goods too quickly at the expense of consumer goods. There is a further danger that too much of the savings may be directed to the development of heavy industry and too little to the development of agriculture, thus killing the goose that lays

[1] T. J. Hughes and D. E. T. Luard, *The Economic Development of Communist China*, 1949–1958 (Oxford 1959), Chapter 7.

the golden eggs. This is a danger that has not been entirely avoided in the last decade of planning.

Since 1959 there has been a series of bad harvests. The fall in agricultural production has been a matter of the gravest concern. There is no doubt that this fall is due largely to natural causes of phenomenal droughts and floods, but there is equally no doubt that loss of production has stemmed from bureaucratic control, over-centralization and over-straining and over-draining of agriculture. That there has been a falling off of enthusiasm is evidenced by the amount of lively discussion in newspapers and magazines of the ideological correctness of the recent extension of privately held small-holdings and the effect that production on such holdings has on such things as the distribution of fertilizer, which is almost entirely night soil and which the peasant tends to hoard for his own plot. It is estimated that small-holdings are producing about 10 per cent of the total annual agricultural output.

The following figures give some idea of both saving and expenditure:

SAVINGS INVESTED IN THE FIRST FIVE YEAR PLAN

Industry	58·2%
Agriculture	7·6% (including water conservancy)
Communications	19·2%
Education and Public Health	7·2%

By 1957, when practically all means of production were state owned and controlled, the budget for that year gives a picture of distribution:

Revenue		*Expenditure*	
Taxation	50%	Economic construction	50%
State enterprises	48%	Defence	20%
Loans	2%	Education and social services	18%
		Administration	9%
		Others	3%

The task of transforming China into a modern industrial state has been conceived and is being carried out in a series of planned periods of five years. The first was from 1953 to 1957; the second, from 1958 to 1962.

The First Five Year Plan aimed to raise the national income by 43 per cent and this, it is claimed, it achieved. Industrial output in that time was increased by 120 per cent and agriculture by 25 per cent. Great emphasis was laid on heavy industry in this period as it was held that, until heavy industry was functioning, technical modernization of agriculture could not be undertaken. The target of the Second Five Year Plan was to raise the national income by a further 50 per cent by 1962. Great emphasis was still placed on industrial development, but an increasing recognition of the importance of agriculture is revealed in that the Plan aimed at a 70 per cent increase in industrial output and 35 per cent increase in agriculture. According to official statistics, agricultural produce was increased by 25 per cent in the first year, 1958, alone—the year of the fantastic 'Great Leap Forward'—but the subsequent years, 1959, 1960 and 1961, of disastrous flood and drought and the break-down of reliable statistics leave considerable doubt as to whether the 1962 target in agriculture was actually reached.

Agricultural Plans

We have already followed the steps of revolutionary change in the structure of rural society after 1949 from a semi-feudal landlord-peasant basis, through land reform, mutual aid team, co-operative to commune. It now remains for us to examine the methods by which the above vast increases were to be secured and the measure of success so far achieved. Two methods are possible: the increase of cultivated area and the increase in output per acre. Both have been extensively followed.

It is estimated that there are between 30 and 35 million hectares of land (75 to 86 million acres) that are worth reclaiming and which, if reclaimed, would add about one third to China's arable land. The aim of the First Five Year Plan was to reclaim 4 million hectares (nearly 10 million acres) and of the Second Five Year Plan to reclaim 5 million hectares (11·3 million acres). Expressed differently, this means an annual increase of about 2 million acres. Inevitably much of the new land brought into cultivation was marginal. In the enthusiasm of the early years of the drive, hillsides above the paddy lands of the south were ploughed—often up and down the slope. Initially there was much unwise cultivation, which was later corrected by contour ploughing, especially in the more vulnerable north and north-west. However, not all the new land brought into cultivation was of poor quality. For example, hundreds of thousands of hectares of dry land in Sinkiang and the north-west have been irrigated and are reported to be very fertile and to be yielding heavy harvests.

In addition to the two Five Year Plans, which are concerned mainly with the fixing of targets, agriculture has also a Twelve Year Plan, formulated in 1956. This plan is concerned mainly with techniques and methods of increasing output per acre. It has zoned the country into three main belts and has set a target of average increased production per acre for each belt, based on the average output in 1955 (See Fig. 41):

	lb. per acre	
	1955	1967
1. N. of the Hwang Ho, Tsinling and Pailung Rivers	990	2,640 (400 *catties* per *mow*)
2. S. of the Hwang Ho, Tsinling and N. of the Hwai	1,373	3,300 (500 *catties* per *mow*)
3. S. of the Hwai, Hwang Ho and Tsinling	2,640	5,280 (800 *catties* per *mow*)

The Twelve Year Plan has also instituted an Eight Point Agricultural Code, which was heralded by one of the many slogans the Chinese love so well, 'Carry out conscientiously the Eight Point Agricultural Code'. These slogans, which often seem naive and even childish to the West and which, when translated, are clumsy, are terse in Chinese. Although great use has been made of slogans in the agricultural and industrial drive, they

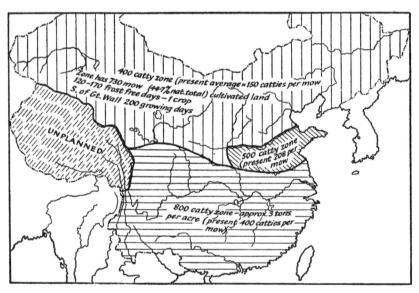

FIG. 41. Agricultural targets for 1967

PLATE XIII: Modern and more efficient means of irrigation make an interesting contrast with the traditional methods seen in Plates III and IV.

PLATE XIV: A stand of Japanese cedar in Hunan. Careful forestry is now being practised here, where in the past there has been serious deforestation.

are by no means an innovation of the present régime. The eight points are: fertilization, deep ploughing, seed improvement, close planting, plant protection, soil improvement, tool production and irrigation.

SOIL IMPROVEMENT AND FERTILIZATION

Over more than 2,000 years a high degree of fertility has been maintained in the heavily worked soil by maturizing night soil and using animal manure, vegetable waste and ashes. The work is laborious and not very efficient. Sr. Guiseppe Regis says 'Given the very low average active fertilizer content of the material used, viz. from 15 to 45 times less than that of chemical fertilizer in the more common compounds, the labour involved in the collection, preparation, transport and spreading of these natural materials is enormous.'[1] New and more efficient ways of maturizing night soil have been devised, and since 1952 much emphasis has been placed on composting, together with the use of mud from ponds, and seaweed.

However, it is to chemical fertilizer that the planners look for big improvements in soil and big increases in yield. Before 1949 very little artificial manure was used. The peak amount before that date was 227,000 tons for the whole country. By 1957 this had risen to 740,000 tons and by 1958 to 1,300,000. The target for 1962 was 10 million tons. Most of what is produced is ammonium sulphate. Thirteen large plants were begun under the First Five Year Plan, the biggest of which is at Kirin. Others are located at Canton, Nanking and Dairen. In addition there are hundreds of smaller fertilizer plants of standard pattern scattered all over the country in the communes. These are usually capable of producing 8,000 to 10,000 tons per annum for use on the local fields.

CLOSE PLANTING AND DEEP PLOUGHING

With the increased use of fertilizer an endeavour has been made to produce heavier crops by close planting. Astronomical yields have been obtained experimentally. For example, in the Yangtze valley where 100,000–200,000 shoots of rice per *mow* were planted, giving a crop of 360 *catties*, 300,000 to 500,000 shoots are said to be planted now, giving a yield of over 1,000 *catties* per *mow*. However, it is difficult to ascertain how far close planting has extended beyond the experimental stage of the agricultural colleges and become common practice. Wherever it is used it must be accompanied by deep ploughing. The old Chinese plough pene-

[1] Guiseppe Regis, 'Developments in Chinese Agriculture', *Far East Trade*, January 1962.

trated to a depth of only 3 to 4 in. The new ploughs being produced turn a furrow of twice this depth. It should be noted how interdependent all these innovations are. Just as close planting is dependent on adequate fertilizers, so deep ploughing relies on increased power. If all that is available by way of power is the light ox of south China, deep ploughing cannot be practised.

SEED SELECTION

One of the main tasks given to the many new agricultural colleges and research stations has been the development of improved strains of all kinds of grain. It is claimed that by 1958 more than one third of China's cultivated land was sowed with the new seed and was partly responsible for the bumper harvest of that year.

Improved seed, requiring a shorter growing period, has enabled rice to be grown successfully even in Manchuria. Double cropping is now possible in the Yangtze valley, thus raising the areas under double crop in China to 7 million acres.

PLANT PROTECTION

Considerable emphasis has been placed on the need for greatly extended pest and disease control. Pests and disease take a toll of as much as one third of the crop in some instances. Rust in wheat, and aphis and red spider in cotton have been particularly destructive, as also have been caterpillars and locusts. All the principle types of modern insecticides and fungicides are being used in increasing quantities. In 1959 these reached a total of 343,000 tons.

TOOL IMPROVEMENT

Mao Tse-tung laid it down that no full scale mechanization of agriculture could be contemplated until the Third Five Year Plan at the earliest, since it must await the full development of heavy industry. By 1959 only 4 per cent of the country's cultivated area, approximately 4 million hectares, had been mechanized. In so far as mechanization with modern machinery has been carried out, it has been largely confined to state farms and tractor stations. In 1960, there were 2,490 state farms, which possessed 28,000 of the 55,000 tractors (in terms of 15 h.p. units) then in existence and an even greater proportion of the country's 4,500 combine harvesters and 7,500 power driven threshers. Progress in the manufacture of agricultural machinery is dealt with in the industrial planning section. Although this machinery is beginning to roll off the lines of the new plant, some idea of the impact which mechanization has made on the country as a whole up to the present can be gathered from the 1959 estimate that only 1 h.p. of machine and generating power was available for 30 units of adult human labour.

But this call for tool improvement in the Eight Point Agricultural Code has not so much the tractor and big scale mechanization in mind as the adaptation and improvement of old, simple tools and the invention of small machines designed and manufactured locally in commune workshops. It is here, probably, that the old, innate genius of the Chinese is showing itself even more clearly than in the new massive, integrated iron and steel plant of the North-East. Modern industrial prowess of the West and Chinese conservatism and instability of the last century have combined to obscure the fact that in the Middle Ages the flow of scientific and technical ideas was from China to Europe, as Dr. J. Needham has been at pains to point out.[1] All the common tools have come under scrutiny. Local groups of farmers, co-operatives and production teams have been encouraged to bring their ingenuity to bear on improving them. Hoes and ploughs have been lightened and improved; simple seed drills invented; and traditional water-wheels made much more efficient by such simple devices as adding fly-wheels. Attempts that have been made to lighten the back-breaking work of rice transplanting have met with apparent success by the invention of several types of simple machine which can be operated by one person and which can be made in the local workshop. Pneumatic tyres and wheels with ball-bearings fitted to carts and wheelbarrows have greatly added to the efficiency of rural transport. All these machines and many others of a similar calibre are produced in the small local workshops and factories. They provide one of many examples of what is meant by 'walking on two legs', yet another of the catchwords or slogans, meaning that the great and the small, the local and the central, the expert and the amateur can work together and be of mutual assistance.

IRRIGATION AND WATER CONSERVANCY

The Five Year Plans include comprehensive national schemes for the control of the great rivers, notably the Hwang Ho, the Hwai and the Yangtze. These vast projects are of a calibre similar to that of TVA and even greater. They are multiple purpose, aiming at flood control, provision of irrigation, the generation of hydro-electric power and the creation of navigable waterways. They all involve great engineering works and the

---

[1] J. Needham, *Science and Civilization in China* (Cambridge 1954), Vols. 1–3.

expenditure of considerable capital. The Hwang Ho programme alone plans the building of fifty dams, one of which, the Sanmen, is as big as the Kariba Dam. Each of these projects will be given detailed consideration later in their appropriate regions.

When the Eight Point Agricultural Code exhorts the farmer or the production team or brigade 'to carry out irrigation conscientiously' it does not envisage the great schemes but multitudinous small local works. It is estimated that in 1952 there were approximately 50 million acres of irrigated land in China. By 1957 this had been raised to more than 80 million acres and in 1958, the 'Year of the Great Leap', it was claimed that a further 58 million acres had been brought under irrigation.

Much of this rapid expansion of irrigated land has been the work of the co-operatives and later of the communes. In the early stages it often consisted simply of the extension of existing irrigation by storage ponds at the head of valleys. With the formation of the co-operatives and then of the production teams and brigades, labour was mobilized and more ambitious projects could be carried out, involving the building of large earth dams and the construction of long aqueducts and large networks of channels. The virtue of these works is that they are carried out with a minimum of capital expenditure, that they impinge very little on big centrally planned schemes and that they can use local labour during slack agricultural periods.

The kind of project undertaken varies as between districts and regions. Northern Shensi, a district in the heart of the loess, has an ambitious scheme, which involves the eventual building of fifty earth dams with the multiple purpose of irrigation, prevention of soil erosion and of soil retention and reclamation. The dams are built across the valleys, and a minimum of stone is used in their construction, it being confined to the base and the spillways. The rest is made of tamped earth, rammed down hard and packed tightly by means of 'frogs', i.e. heavy round stones, raised by teams of eight to ten men and then dropped. It is reckoned that each of the lakes caused by the dams will silt up in five or more years and will then contribute many acres of valuable arable land. As each lake silts up fresh earth dams will be built elsewhere. The scheme is linked with one of afforestation throughout the district. If such projects as this become common in the north-west, they would contribute considerably to the solution of 'China's Sorrow'.

Farther south in a district in Hupeh another irrigation system, which has been dubbed 'The Water Melon Scheme', has been evolved. A stream is dammed at various spots, reservoirs are formed and irrigation channels are led off to the fields and to storage ponds against the dry winter when the flow of the stream is low. Presumably it is this last feature that has given the scheme its name. The project, which is reported to be working satisfactorily, was conceived and carried out entirely by local labour, using only the simplest of implements. It is stated that the surveyor's levels used were light bamboo tubes floated on rice bowls filled to the brim with water.

A feature of the irrigation programme on the North China Plain has been the sinking of thousands of new wells.

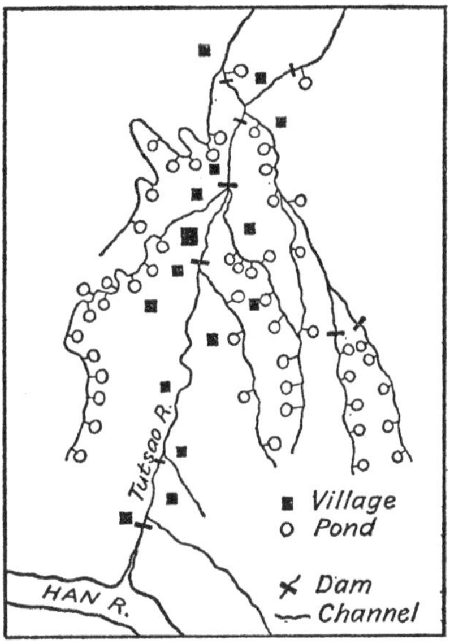

FIG. 42. Water melon system of irrigation, Tutsao River, Hupeh

AFFORESTATION

In a country whose climate over vast areas is favourable to tree growth and whose natural vegetational cover should be forest and woodland, it is startling to find that only about 5 per cent of its total area is in fact forested. Centuries of ruthless clearing and of rigorous extension of ploughlands have largely been responsible. There is a Chinese saying which runs 'He who plants trees does not enjoy the shade but he who comes later'. The low value set on trees and tree planting, together with the ever increasing demand for fuel, either as wood or charcoal, has led to widespread and disastrous deforestation.

The Twelve Year Plan has as its object in this field first the education of the peasant masses to the great value to themselves of forest and woodland for the prevention of soil erosion and for the control of flood and drought. The full co-operation of the peasants must be enlisted in any big afforestation project, as it is on their labour and their understanding that the planners have to rely. In recent decades many local attempts have been made at afforestation and all have come to grief, partly through unsettled and chaotic political conditions, which have rendered proper conservation impossible, and partly also through ignorance of the value of what was being attempted.

Here again the problem has been tackled at both the national and the local levels, again 'walking on two legs'. In the north the campaign to deal with the floods of the Hwang Ho, the dust storms and the prevention of a dust bowl has been launched with a slogan, 'Taming the "Yellow Dragon".' It is planned to plant a shelter belt of trees right across the dry north-west, nearly 1,000 miles in length and having an average width of one mile. It is claimed that much of this has already been planted and established, 700,000 peasants having worked on it in Kansu alone in 1958. The work of conservation and protection of the saplings against sandstorms until strong enough to stand on their own has been largely the responsibility of the communes of each district. Once established, moisture in the sub-soil of the dry regions appears to be sufficient to support healthy tree growth. Records from forestry stations in the north-west state that the shelter belts have cut down wind speeds from an average of 38 m.p.h. to 15 m.p.h. and are being successful in the control of sand movement in the Ordos and the loess region. Wherever possible fast growing varieties (fir, larch, poplar and camphor) are grown as they are valuable later as timber.

Between 1949 and 1958 it is reported that 5,400 sq. miles have been planted with trees, 12,700 sq. miles sown with grass and shrub and 15,000 miles of wind break established.

RAW MATERIALS FOR INDUSTRY

Since four-fifths of the raw material used in light industry in China comes from the farms, it is not surprising that the increase of production of these raw materials has been the subject of a good deal of attention in the Five Year Plans. Natural calamities such as flood and drought result immediately in bad harvests and a fall in supply of raw materials which, in its turn, leads to a slump in light industry. While in the earlier planning years heavy industry occupied the centre of the stage, the series of bad harvests in 1959, 1960 and 1961 have brought into prominence the important role that agriculture has to play in industrial development.

Cotton is the most important of the industrial crops. The chief producing regions remain the same as in the pre-1949 period, but by better seed selection the growing area has been extended and the quality improved. Hopei, Honan, Shantung, Hupeh, Kiangsi, Szechwan, Shansi and the north-west all now grow cotton of 1 in. staple or more. Although production has increased, it has not always reached the target for the year.

		1,000 tons					Planned
Peak Year	1943	1949	1951	1952	1953	1957	1962
	1,000	701	936	1,394	1,503	1,650	2,400

A species of wild hemp, capable of resisting wind, drought and cold has been found. Moreover it can be grown in alkaline soil and is thus eminently suitable for growth in the north-west. The growth of flax has been increased in the North-East and some jute is being grown in the south-east.

The cultivation of both cane and beet sugar has been stepped up very rapidly. There have been big cane developments in Kwangtung and southern Yunnan and extended growth of beet on the Sungari plain. The pre-1949 peak production was 414,000 tons. This fell to 199,000 tons in 1949 but rose steadily and rapidly to 1,300,000 tons in 1958. A total of 2,400,000 tons was planned for 1962. The First Five Year Plan made provision for the erection of 20 large and 81 small plants to process cane and beet. A big, fully-integrated plant, 'The Sugar Cane Chemical Complex', has been built at Kiangmen, near Canton. In addition to processing the sugar itself, it produces bagasse for paper pulp, etc., molasses fermented to make alcohol, raw materials for nylon and plastics and also dry-ice and yeast. Many communes both in the south and north run small handicraft sugar factories, producing about 5 tons per day.

One of the most urgently needed agricultural raw materials for industrial purposes is oil seed. Peasants everywhere have been encouraged to grow soya, peanut, rape, sesamum, linseed, sunflower and castor oil, and have been urged to use every bit of waste and vacant land, such as the verges of highways and embankments, with a result that production has greatly increased:

Peanuts	5,000,000 tons (1958 figures)
Sesame	550,000 ,,
Rapeseed	1,384,000 ,,
Soya	12,000,000 ,,

Sericulture, which suffered so heavily during the Sino-Japanese War and also from the competition of synthetic products, has not been restored to its pre-war level. It has been encouraged in Kwangtung and Kwangsi, where the climate is such that 7 to 8 hatchings can be achieved in a year. Greater care is now being given to all the processes, including the growth of mulberry trees. Successful experiments have been carried out in the growth of the castor oil plant as food for the silkworm. It is claimed that castor oil grows faster and on poorer soils than mulberry and also needs less labour.

Some state mechanized farms in Kwangtung and Hainan are experimenting and specializing in the production of tropical crops, notably rubber, copra, cacao and coffee, palm oil, quinine and rattan.

Industrial Planning

We have seen that during the past half-century and more, industrial and commercial development in China had been mainly along the coast and had been concentrated heavily in such ports as Shanghai, Tientsin and Canton. This was due largely to two factors: (1) the absence of good internal communications, with the exception of the Yangtze waterway up to Ichang and (2) the reliance on largely foreign capital, which, being concerned mainly with international trade, tended naturally to be invested along the ocean periphery. In China Proper there was no really big centre of heavy industry, although a good deal of lighter iron and steel manufacture was carried on in Shanghai. The North-East (Manchuria) was the only area of heavy industry and this was almost entirely in the hands of Japanese industrialists. Here modern integrated iron and steel plant, located in Anshan and Shenyang, was responsible for a very considerable production.

During the Sino-Japanese War, and more especially during the civil war which followed between 1946 and 1949, industry throughout China suffered severe deterioration. By 1949 it had slumped to its lowest level in every branch. It is necessary, therefore, when comparing achievement before and after 'liberation' to look to the peak period before 1949 rather than to 1949 itself. Industry in the North-East suffered particularly seriously, for not only did it suffer spoilation at the hands of Soviet forces, who came in in 1946 and carried off the greater part of the iron and steel plant, but production came virtually to a standstill as Communist and Kuomintang forces fought for control of the region.

As with agriculture, so with industry, the years between 1949 and 1952 were spent in recovery and rehabilitation before embarking on the stupendous task of converting the country into a modern industrial state. Briefly the policy was one of dispersion and expansion. By dispersion was meant a much wider distribution of industry away from its concentration along the littoral, thus developing the whole country. The existing coastal concentration was associated in the early planners' minds with foreign domination, and in the first flush of revolution in 1949 this distribution came under heavy criticism. For a while there was a movement actually to transfer established industry, particularly that of Shanghai, to places inland. For a good many of the years of the First Five Year Plan distribution in depth was the avowed objective, but while this still remains the aim, maturer thought has given this peripheral development a more generous assessment.

'During the Second Five-Year Plan period we must continue the construction of the industrial bases in Central China and Inner Mongolia with the iron and steel industrial bases in South-west China, North-west China and the area around Sanmen Gorge, with the iron and steel industry and large type hydro-electric power stations as their core; carry on with the building of oil and non-ferrous metal industries in Sinkiang; and intensify geological work in Tibet in order to prepare the way for its industrial development . . . We may say that the existing industrial bases in the areas near the coast are the starting point of the industrialization of our country.'[1]

Behind the move for dispersion were the rational ideas of the creation and development of a wide national market and of a much fuller use of local resources to meet local needs, thereby economizing in transport and conserving it for more pressing needs. This wider distribution of necessity applied mainly to medium and light industry. The location of heavy industry is largely dictated by the siting of its raw materials and the sources of its power. As the preponderance of these is found in the north, it is not surprising to find that of the five centres selected for fully integrated modern heavy iron and steel plant, four are in the north, namely the North-East, Paotow, Taiyuan and Tientsin, while Wuhan is the chosen centre farther south.

IRON AND STEEL

The intensive geological field survey work which has been carried out since 1949 has disclosed vast new coal and iron ore resources. The great proportion of coal deposits is still located in the north, as the following figures for the distribution of proved coal reserves in 1957 show:

	%
Shansi	31·81
Inner Mongolia	25·57
Honan	6·97
North-East	6·45
Hopei	5·54
Shensi	5·25
Sinkiang	3·62

Early post 1949 estimations of iron ore reserves stood at 4,338 million tons, of which 3,189 were in the North-East. In 1957 the geological survey reported a proved reserve of 7,900 million tons, most of which has over 40 per cent iron content. Even newer, richer finds are claimed in Shansi

[1] 'Report on the Proposals for the Second Five-year Plan for Development of the National Economy' by Chou En-lai, 16th September 1958.

and Honan, where the reserves are now estimated to be 7,000 million tons. Rich sedimentary deposits (40–60 per cent) are reported in Kwangtung and Hainan. Further finds are recorded in Chahar (207 million tons), Suiyuan (157 million tons) and Hupeh (173 million tons).

The North-East was an obvious choice for the development of heavy industry since it had already been established here for several decades. In spite of the destruction which had taken place in the post-war years, industry was rapidly rehabilitated. Plant destroyed or removed was replaced by the most up-to-date machinery from USSR. By 1952 the North-East was producing 50 per cent of the nation's coal, 80 per cent of its power and 90 per cent of its iron and steel. Since that date enormous expansion has been carried out in the network of North-East industrial towns. Shenyand (Mukden) is the focal point for the whole area. There, and also at Tsitsihar and Changchun, machine-tools are the main product. Anshan is the main centre for iron and steel, closely followed by Penchi, Harbin, and Dairen. Two great open coal cuts are at Fuhsin and Fushun and the main centre of the chemical industry is at Kirin.

A great and rapid development has taken place at Paotow on the coal-fields of Inner Mongolia, Shansi and Shensi and on the newly discovered iron resources in that district. Four blast furnaces and five open hearth furnaces have been built, having an annual output capacity of 3–5 million tons of pig-iron, 3 million tons of steel and 2 million tons of rolled steel. Similar but not such vast developments in heavy industry have taken place at Taiyuan and Shichingshan in Shansi and at Tientsin in Hopei.

At Wuhan, although not exactly on the same site as the old Hanyang Iron and Steel Works, the moribund heavy industry has been resuscitated in no uncertain manner. Integrated plant of the most modern type, including two blast furnaces, each of 1,500 cu. metres capacity, has been erected. Wuhan draws its raw materials, as formerly, from Pingsiang (coking coal), Tayeh (iron ore) and Shihweiyao (limestone).

In view of the vital importance of steel for the construction of all the railways, bridges, factories, transport and buildings called for in both the First and Second Five Year Plans, priority has been given to the development of heavy industry.

Peak steel production before 1949 had been 923,000 tons. This fell to only 158,000 tons in 1949. In the rehabilitation years it rose again to 1,349,000 tons in 1952. During the First Five Year Plan production rose rapidly and steadily, each year's target being overshot. In the last year or two of this period there was much consideration and discussion of the need for greater decentralization of direction of all industry, especially iron and

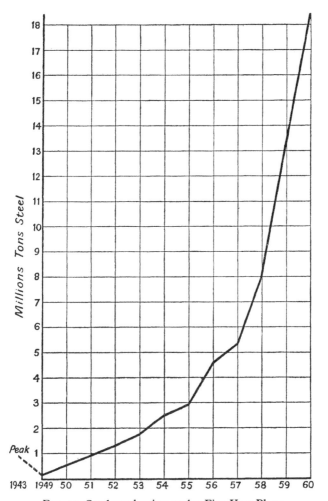

FIG. 43. Steel production under Five Year Plans

steel, in order to speed up production, and in one year the output was raised from 5·25 to 10·7 million tons. As we have seen (p. 168), in the enthusiasm of the 'Great Leap Forward', backyard furnaces were introduced. They resulted in a serious diversion of rural labour from agriculture at a moment when big crops were awaiting harvesting.

The hour of these small, backyard furnaces was short. They have quickly fallen out of operation and have been replaced by local medium-sized modern plant, capable of turning out 500–1,000 tons a week. About half the nation's pig-iron is produced in this way. Some idea of the wide distribution of such plant is gleaned from the fact that the Lhasa Iron and Steel Plant produced its first pig-iron in October 1960.

The acceleration of steel production, started in 1958, has continued at undiminished speed. In 1960, 18,450,000 were produced, thereby outstripping France's output of 17,200,000 tons. H. W. A. Waring, writing on the subject said 'Chinese steel production is currently running at an annual rate of 18 million ingot tons. It is increasing at a rate of 4 to 5 million tons a year and by 1962 may well top the British figure which is currently around 25 million tons'.[1] The UN forecast of Chinese steel production for 1972 is 52 million tons, while the Chinese themselves estimate that they will pass the 70 million ton mark by 1970. Even if this great increase is achieved it will mean a consumption of only about 150 lb. per head, since, according to estimates, China's population will be 1,000 million by 1970. Comparable figures for the USA are 1,270 lb., the UK 875 lb. and the USSR 560 lb. per head.

Coal production has increased at a speed equal to that of steel. Peak output before 1949 was 61·9 million tons. This fell to 30·9 million tons in 1949. The First Five Year Plan fixed a target of 113 million tons for 1957. This was exceeded by 11 million tons. 190–210 million tons were planned for 1962, the last year of the Second Five Year Plan, but this was exceeded in the 'Great Leap Forward' in 1958. The reported production in 1962 was 425 million tons. The greatest output is still in the North-East, but there has also been a big rise in production throughout the country. Many new mines have been opened, old mines modernized and a great deal of the latest coal cutting machinery installed.

Medium and light industry has, to a large extent, had to wait upon the development of heavy industry, both for the manufacture of the initial tools and machinery and also for its own raw material. This has been particularly true of the car industry, which is very dependent upon the production of sheet steel.

The first modern motor works was planned under the First Five Year Plan and opened at Changchun, North-East, in 1956. Production in 1957 was not very large, owing to the shortage of sheet steel, for it amounted to only 7,500 four-ton lorries. Under the Second Five Year Plan production has been vastly increased. Six basic types of car are made, namely, heavy and medium lorries, large, medium and light touring cars and a general purposes farm car of the 'Land Rover' class. Output from this one plant is planned to reach 300,000 a year in 1970. New plant has also been built at Nanking for the production of 1½-ton lorries. This has a capacity of

[1] H. W. A. Waring, *Steel Review*, British Iron and Steel Federation 1961.

100,000 lorries per annum, and should have been in full production in 1962. In 1959 No 1 Tractor Works was opened at Loyang. This factory had an initial annual capacity of 15,000 tractors of 54 h.p. Four other major tractor plants at Tientsin, Shenyang, Anshan and Changchun have since been built and all are now in full production, giving a total annual output of 100,000 tractors. The Loyang factory has complete Soviet plant installed by USSR experts, but it is now run entirely by Chinese personnel, who received their training in Russia.

In 1956 China was able to produce only about 20 per cent of the metals and alloys used in the car industry. By 1959 this percentage had been raised to 90.

Iron and steel works, car factories, machine-tool shops, chemical works and textile mills are used a great deal as training centres. Workers from the communes are taken into the factories in large numbers for some months where they work and learn, returning to their own local workshops to teach others. For example, by 1960 the Dairen Chemical Co. had given some 8,000 commune members training in the production of chemical fertilizers. The importance of the commune workshops and factories can be assessed from the fact that they account for 10 per cent of the country's industrial production. The urban communes' plant and workshops are engaged mainly in the production of accessories and semi-finished articles for the big integrated works.

Probably more important than all the vast material development which has taken place in the last decade is the rapidity with which such big numbers of Chinese, formerly quite unskilled, have learnt the skills and technical know-how of modern industry in all its complications. A striking feature of the suburbs of modern Peking is the large number of great technical institutes that have sprung up to meet this need for knowledge and skill. And what is true in this respect of Peking is true also of every large centre throughout the country. Schools and particularly technical colleges have received a high place in priorities. The growth in the output of cement reflects in some degree the great building boom which has taken place.

OIL

Weakest of China's power resources is oil. However, recent surveys and discoveries have raised hopes that China may yet become one of the world's greatest producers.

Since 1949 the old Kansu oilfield has been extended and production stepped up steeply. New and most promising fields were discovered in

Dzungaria in 1955 at Karamai and Tushantze and in the Tsaidam at Manrai and Ta Tsaidam. The Karamai field holds the largest reserves and now has sixty-two wells. In Szechwan a large field was discovered in 1958 on the Chialing River in both the Lower Jurassic and Triassic. The presence of natural gas here had been known for many centuries and the hope of oil has inspired intensive search in recent decades, but it was not until 1958 that oil was actually struck. This discovery promises to make a great difference to the economy of the province. Recent claims are also made of big discoveries in Yunnan and Kweichow and in the area between the Liao and Sungari Rivers in the North-East. Hopes are held of big finds in the Tarim, Turfan, Tibet and Kwangsi, but these areas have not yet been adequately surveyed. The oil shales of Fushun in the North-East have continued to be worked energetically.

In 1949 only four oil and two natural gas fields were being operated. By 1960 these figures had risen to 32 and 18 respectively. Production has risen spectacularly as the following figures show, although in terms of world production they are not very significant.

Crude Oil Production ('ooo tons)

	1949	1950	1951	1952	1953	1954	1955	1956	1957	1958	1959
National	120	200	305	436	622	789	966	1,163	1,460	—	3,700
Imported from the USSR	—	—	—	607	—	—	948	1,090	1,757	2,507	—

Refineries have been built at Lanchow, Yumen, Karamai and Lenghu (Tsaidam), and others are being built in Szechwan and in the south. Imports of oil from the USSR have also risen steeply as the production of cars has increased and roads have been built. It is instructive to note the correlation of these three, the production of oil and cars and the construction of roads.

HYDRO-ELECTRIC POWER

Apart from the hydro-electric power developed by the Japanese in the North-East at Fengman, China's great water power potential, now estimated at 544 million kW, has remained undeveloped until recently. It is estimated that the capital cost of development in China is very low as compared with that of Europe and America, and is only about one and a half times the cost of building thermal plant. This, it is maintained, is due partly to favourable topography. China's great rivers flow from the high west to the low east. They have big falls and narrow gorges in the west and centre, e.g. Liuchia and Sanmen on the Hwang Ho and the Gorges between Wanhsien and Ichang on the Yangtze. In the east the great volume of flow can be utilized. In addition there is great water power potential in the south-west, where coal reserves are least.

Very ambitious plans for the development of this resource were formulated in the First and Second Five Year Plans. China has received a great deal of help in this field from the USSR, both in surveying and in building power stations, especially in the early years when all the plant used was imported from Russia. China has now learned to produce and build her own plant, e.g. the 60,000 kW plant at Shangyiukiang, in Kiangsi, its four 15,000 kW generators having been built in Shanghai and Harbin.[1]

Most of the big hydro-electric projects already completed or now being undertaken are integral parts of great water conservancy schemes. The Hwang Ho Conservancy Scheme includes the building of forty-six 'steps' or dams, which, when completed, will generate more than 2,300,000 kW. Included in this are the two great dams at Liuchia and Sanmen, each of which has a capacity of 1,000,000 kW.

In the North-East the greatest station is at Fengman. This great dam was built by the Japanese but was damaged in the war and fell into disrepair. Its reconstruction was completed in 1957 and it is now generating 567,000 kW. Big joint Chinese-USSR schemes are in hand on the Amur, where the potential is estimated at 13 to 20 million kW. Other big stations built under the Second Five Year Plan are:

> Kwantung, Yungting River (near Peking)
> Kungki River, Szechwan (100,000 kW)
> Sinan River, Chekiang (580,000 kW)
> Ili River, Yunnan (360,000 kW)

It is estimated that when the 'Three Gorges' station on the Yangtze in Szechwan is built it will generate 20 million kW.

The Hwai Ho Conservancy scheme includes the construction of twenty-seven reservoirs. Some of these have now been completed, including those at Meishan and Futzeling. These are medium sized stations, typical of many that are being built. Mention should also be made of projects in Sinkiang, where attempts are being made to harness the snow-fields and glaciers of the Astin Tagh (Altyn Tagh) for both irrigation and power.

In the rural districts encouragement is given to the communes to

[1] *Far East Trade*, May 1958.

develop small hydro-electric stations to provide power for irrigation purposes.

It is estimated that hydro-electric power now represents 28 per cent of the total power generated in China.

Production under the First and Second Five Year Plans

Product	Pre-1949 Peak	1949	1952	1st 5 Y.P.	2nd 5 Y.P.	
				1957[1] (planned)	1958	1962 (planned)
Grain (incl. pulse) (million tons)	139	109	154	182	185	250
Pigs (million)	78·5		89·8[2]	138·3		250
Sheep and Goats (million)	5·96	1·8	7·26	15·9 (19·0)	53·5 (1960)	40–43
Sugar (thousand tons)	414	199	450	864	1,300	2,400
Raw Cotton (thousand tons)	850		1,300	1,650		2,400
Cotton yarn (million bales)	2·45		3·62	5·0 (4·65)		8–9
Oil-bearing crops (million tons)	17·37[3]	7·41	13·25	13·82	20·00	
Chemical fertilizer (thousand tons)	227		194	570 (740)	1,300	10,000
Coal (million tons)	61·9	30·9	63·5	113 (124)	425 (1960)	190–210
Pig-iron (million tons)	1·8	0·21	1·9	4·7 (5·86)	27·5 (1960)	
Steel (million tons)	0·923	0·16	1·35	4·12 (5·24)	18·45 (1960)	10·5–12
Machine-tools (pieces)	5,390	1,582	13,734	37,192	80,000	

[1] Figures in brackets show production actually achieved.

[2] There was a marked decline in 1956, probably owing to doubt as to compensation and ownership of animals in co-operatives; figures rose again when individual ownership was assured.

[3] Composed of peak production of soya (1936), peanuts (1933), rapeseed (1934) sesame seed (1933).

Production under the First and Second Five Year Plans—continued

Product	Pre-1949 Peak	1949	1952	1st 5 Y.P. 1957[1] (planned)	2nd 5 Y.P. 1958	2nd 5 Y.P. 1962 (planned)
Crude oil (million tons)	0·32	0·12	0·44	2·0 (1·42)		5–6
Electric power (milliard kW)	5·96	1·8	7·26	15·9 (19·1)	53·5 (1960)	40–43
Cement (million tons)	2·4		2·6	6·0 (4·65)		12·5–14·5
Machine-made paper (million tons)	0·165	0·108	0·372	0·913	1·300	
Railways (miles)	15,600		15,100	17,600		23,000
Passengers (million)	265	157	164	313	330	

The Development of Communications and Transport

Until the end of the nineteenth century communications and transport in China were essentially the same as in the days of the Han dynasty. The Court of the Ching dynasty relied upon canals and navigable rivers in the main for the delivery of the imperial tribute grain just as their T'ang predecessors had done. Carts and pack animals were used in the north over roads that were nothing more than earth tracks. In the south the coolie with his carrying pole or squeaking wheelbarrow plodded his weary way over paths, which in their better parts were paved with stone slabs and in their worse sections were deeply rutted traps for the unwary walker.

The efficiency of communications fluctuated with the political state of the time. In times of vigorous central government at the advent of a dynasty, when the empire was far-flung, speedy contact with outlying parts was essential and was maintained by posting by horse in the north, and largely by runner in the south. It was the efficiency of this which so excited Marco Polo's admiration when Mongol power was at its height in the Yuan dynasty. Friar John wrote of the speed with which he travelled when called to the court of Kuyak Khan:

[1] Figures in brackets show production actually achieved.

'Then we came to the land of the Mongols, whom we call Tartars. Through the Tartars' land we continued to travel for a space of three weeks, riding always hastily and with speed . . . Often changing our mounts, for there was no lack of horses, we rode swiftly and without intermission, as fast as our steeds could trot.'[1]

As dynasties decayed, so did their lines of communication. By the middle of the nineteenth century and after the devastation of the Taiping Rebellion (1850–64) the decadent state of the Ching dynasty was reflected in the chaotic state of its communications. It was then that the impact of the West began to make itself felt. Following the defeat of China in the two wars of 1840–42 and 1856–60, the opening of the Treaty Ports and the granting of concessions, foreign capital began to flow into China, and with it came the development of modern communications.

Railways

In 1876 a first short railway was built between Shanghai and Woosung, but it aroused so much local opposition that the rails were taken up and removed to Taiwan. This hostility (through regard for *feng shui*) persisted throughout the remainder of the nineteenth century and very little building was done. In 1886 government permission was given for the construction of a railway along the edge of the Tai Hang Shan to serve the coal mines of the newly started Kailan Mining Corporation.

The first decade of the twentieth century saw a minor boom in railway building and by 1916 China had 3,431 miles (5,490 km) of track. During this time the two main north-south lines from Peking to Shanghai and Hankow respectively were completed. The Peking-Hankow line, via Paoting and Chengchow, where a bridge carried it across the Hwang-ho, was begun in 1898 and opened in 1905. Built largely with French and Belgian capital, it is 755 miles (1,214 km.) long. This was extended, using Chinese capital, by a line from Wuchang to Changsha which was opened in 1918. The two lines were connected by ferry only across the Yangtze, which is nearly a mile wide at that point.

The second north-south line from Peking to Pukow (Nanking) was begun in 1908 and was not finished until 1918. This was a joint Anglo-German venture. Here again a ferry over the river from Pukow to Nanking made connection with the Nanking-Shanghai line, which had been completed in 1908 and extended to Hangchow and Ningpo in 1912.

---

[1] *The Journey of Friar John of Pian de Carpini to the Court of Kuyak Khan 1245–1247*, Manuel Komroff (Ed.) (London 1929).

The line from Peking to Mukden via Tientsin and Shanhaikwan, which was started in 1880, was not completed until 1907. The capital for this was supplied by the Chinese government and a British loan. The main east-west line from Haichow via Hsuchow, Kaifeng, Chengchow, Loyang to Sian was built between 1905 and 1916. This is known as the Lunghai Railway and was built with French capital. It crosses the rich North China Plain and provides a much needed east-west line of communication, which the meandering, unnavigable Hwang Ho fails to give. No such east-west line has been built through the more fertile and more populous Yangtze valley because the river here has provided excellent transport facilities. For similar reasons Liuchow (Liuyang) has not yet been linked with Canton along the Si Kiang.

Other lines worthy of note, built during these early days of railway development in China Proper, were the Peking-Paotow Railway, which carried one line to the pass and gate of Changkiakow (Kalgan) and another westward to Tatung. Begun in 1905, it reached Kalgan in 1909, Tatung in 1915 and Paotow in 1923. A remarkable narrow gauge (1 metre) railway was built between 1901 and 1910 through the very difficult country from Haiphong (Tongking) to Yunnanfu. All the above railways were single track.

In the North-East (Manchuria) railway development came rather earlier than in China itself. In this development Russia and Japan played leading roles. In 1897 Russia began to build a railway network with a 5 ft. gauge in the north with a main line from Manchouli to Vladivostok, via Tsitsihar and Harbin, thus short-circuiting the Siberian line, which followed the international boundary. This northern development was known as the Tung Ching or Chinese Eastern Railway and was opened in 1903. Farther south the Russians built another system, the South Manchurian or Nan-Man Railway extending from Changchun to Dairen and Port Arthur. In 1904 when Japan defeated Russia, she took over this Nan-Man network and changed it from broad gauge (5 ft.) to standard (4 ft 8½ in.), double tracking some of it. Later, in 1932, when Japan annexed Manchuria, calling it Manchukuo, the whole railway system came under Japanese control.

With the disorder and political instability which followed the Revolution of 1911 and the formation of the Republic, and which continued with little interruption until 1949, the railways of China fell on bad times. There was a period of active construction between 1928 and 1936, in which time railway mileage in China Proper was increased from 4,266 miles to 6,538 miles. The Wuchang-Changsha line was extended southward to meet the

Kowloon-Canton line and thus Peking was linked with Canton. Thereafter little was done either in building or in repair work. The country was never under effective central united control, and consequently no co-ordinated railway policy was implemented until a new era opened in 1949.

We have seen that the Communist Government planned a vast and rapid expansion of industry and a re-distribution and dispersion of that industry over the whole country, creating new widespread markets. A *sine qua non* of such plans was an equally big and quick provision of communications.

The government has used about one fifth of the total state capital investment in both its First and Second Five Year Plans for this purpose:

Total planned state investment in Communications

1953 %	1960 %
13·3 Rail	15·5 Rail
3·8 Other	5·4 Other
17·1	20·9

Although other forms of transport have not been ignored, the mode chosen for main development has been the railway, as the above figures show. In spite of all the modern development of the internal combustion engine, the railway still remains the fastest and cheapest means of conveying bulk and heavy goods long distances overland, at least in countries not yet fully developed economically. It is the means of transport best suited to China's present needs.

During the First Five Year Plan (1953–8) there was little railway building in the populous eastern part of the country. The industrial North-East already had a fairly well developed network which was able to carry the increasing production of that area without too great a strain. Apart from two lines of strategic importance, one from Yungtan to Amoy on the Formosa Strait and the other from Liuyang to Munankwan on the Vietnam border, practically all railway construction was in the west. It was not until after 1958 that the programme changed and building in the east was stepped up.

The most spectacular and most important new railway construction has been the extension of the Lunghai Railway, which runs from Hsuchow to

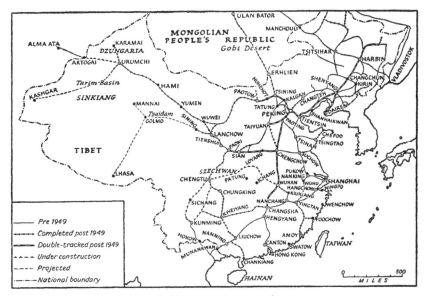

FIG. 44. Railway development since 1949

Sian. In 1950 this line reached only as far as Tienshui. It has now been extended through Lanchow, Wuwei, Yumen, Hami to Urumchi, whence it is to continue to the USSR border, a total distance of 1,750 miles. Here it will meet the Russian line already constructed across Turkestan from Alma Ata and thus will constitute a direct link with USSR.

The Lunghai Railway has both administrative and economic value. It brings the provinces of Sinkiang and Dzungaria more firmly and intimately within the People's Republic and under central control. In times past, even when communications were brilliantly developed as in the Yuan dynasty, these outlying regions were never really an integral part of China. From an economic point of view this railway connects the promising oil-field of Karamai in the Dzungaria basin and those of the Tarim basin and Turfan with the heart of the country, and serves also the oilfields of western Kansu. The line also provides an outlet for the agricultural produce and raw materials, especially cotton, now coming from the newly developed irrigated areas of Sinkiang around the edge of the Taklamakan desert—a development foreseen and advocated more than fifty years ago by Ellsworth Huntington.[1] Urumchi is to be the hub of a railway network linking Karamai, Kashgar, the Turfan and the USSR.

[1] Ellsworth Huntington, *Pulse of Asia* (Boston 1910).

195

The construction of the Lunghai Railway and all the others in the desert and mountainous west has been fraught with great difficulty. Sand dunes and shifting sands in the desert regions have constituted a real problem, which has been partly solved by researches of the Desert Control Station of the Chinese Academy Institute of Forestry and Pedology. It has been found that lower sand layers which contain 2 per cent or more of moisture are capable of supporting saplings, and so many of the dunes can be anchored. The crossing of the Tienshan has been particularly difficult, involving a great deal of tunnelling through heavily jointed ancient quartzite. Soviet help has been lavishly used both in surveying and in construction. The most modern track-laying techniques have been used in conjunction with mass human labour, resulting in amazingly quick but not always wise development.

Another line of great interest is that which has been built between Chining and Ulan Bator (Urga) across the Gobi Desert. Just as the Chinese Eastern Railway from Harbin to Manchouli short-circuited the northern loop of the Siberian Railway to Vladivostok, so this new line further shortens the railway route from Moscow to Peking by some 700 miles. Its commercial importance lies in the direct connection it gives to one of the rapidly developing industrial areas of Siberian USSR. The Chinese section of this railroad between Chining and Erhlien is only 210 miles. The desert section across the Gobi is the work of USSR engineers. The Peking-Paotow line has been extended following the Hwang-ho upstream to Lanchow.

A new railway, the construction of which captured the imagination of the Chinese public, is the one linking Chengtu and the Szechwan basin with Paochi on the Wei River and so with the east-west Lunghai Railway. It is 421 miles long and follows almost exactly the ancient courier 'trestle road' of the Later Han dynasty (AD 25–220).[1] This route was surveyed and the building actually started under the Kuomintang Government between 1936 and 1948, but its completion was left to the present government. It has attracted much attention partly because of engineering difficulties encountered in surmounting—and largely penetrating—the formidable slopes of the Tsinling and Ta Pa Shan. It is reported that about one third of the track in that sector consists of tunnels. A new railway also links Chengtu with Chungking.

Two other new lines lead out of Szechwan. One runs from Chungking to Kweiyang. When complete it will link with Liuchow on the Si Kiang. The other, which is under construction, runs from Chengtu to Kunming.

[1] H. J. Wiens, 'The Shu Tao or Road to Szechwan', *The Geographical Review*, Vol. 39.

A third line is projected and is to go also from Chengtu to Kunming via Sichang. All these three are through very difficult terrain.

A line has been completed from Lanchow to Sining and is to be continued to Gormo and Mannai in the Tsaidam where considerable industrial and agricultural development is taking place. Eventually it is intended to extend this line from Golmo to Lhasa.

After 1958 railway policy underwent a change and much more attention has been given to the needs of the east. Existing tracks have been improved. The entire stretch from north of Harbin to Changsha, via Peking and Wuhan, is now double-tracked, as also is the Lunghai line from Chengchow to Paochi and the Peking line to Paotow. This double tracking has helped to relieve the congestion which became serious with the great increase in industrial activity in the east, especially from 1958 onward. The problem of keeping pace with increasing production remains acute, and rail transport, in spite of all that has been done, is still inadequate. The north-south route from Peking to Canton has been greatly improved and a day's time saved by the building of the Wuhan Bridge over the Yangtze in 1957. The Yangtze was also bridged at Chungking in 1959 and a further bridge is contemplated between Pukow and Nanking, which will then give uninterrupted rail access to Shanghai.

Under the Second Five Year Plan three railways were scheduled for electrification: Paochi-Chengtu, Peking-Paotow, and Tatung-Paotow.

A unique feature of the railway construction programme is the part which the local communities are playing. Distinct from the big national lines, which are planned and built under central government direction, many branch or feeder lines serving mines, forestry work and local needs, are being built and managed by the local authorities or communes. Local labour and, as far as possible, local material are used. All kinds of improvization and make-shifts are employed; for instance rails made from local cast iron or even wooden rails are used and on these refitted locomotives or adapted lorries are run. All are intended as temporary expedients until proper modern means are available. 'Anything that goes, goes' is the slogan here. About 6,000 miles of this kind of feeder line are planned by 1962.

Railway Development

	1952	1953	1954	1955	1956	1957	1958	1959	1960
Trunk lines (thousand miles)	15·1	15·4	15·9	16·8	18·2	18·7	19·5	21·3	24·5
Freight (million tons)	132	160	193	193	246	256	381	542	720
Locos (annual production)		21	52	98	184	267	350	555	835

Roads

Although road building has received little attention in comparison with rail, it has not been entirely neglected. It is estimated that in 1960 there were 156,200 miles (250,000 km.) of surfaced road and 169,000 miles (270,000 km.) of unsurfaced road available for wheeled use only in the dry season. A great deal of the unsurfaced roads are local feeder ways and play a big part in the production drive. They are used by light-wheeled traffic, such as carts with pneumatic tyres and light barrows with bicycle wheels and are eliminating the coolie pole, the most uneconomic and expensive form of transport, as Sun Yat Sen was at pains to point out in his *San Ming Chu Ih.*

	1953	1954	1955	1956	1957	1958	1960
Roads (annual new construction in thousand miles)	5·6	2·1	12·6	40·3	14·4	94·0[1]	40·0[2]

Air Transport

Civil air transport, most flexible of all means, developed rapidly but precariously after the Sino-Japanese War (1936–45). It suffered a severe setback at the time of the Communist victory in 1949 by the loss of most of the aircraft. Consequently its use has been reserved, until recently, for only most pressing services—the rapid transportation of essential personnel and equipment. It is entirely under central control, i.e. there are no private lines.

All the main cities are linked by air, and in recent years there has been a wide and rapid expansion of services. China is beginning to produce her own aircraft in quantity.

The following new air lines were opened in 1959:

Lanchow to Hsining	Urumchi to Karamai
Hofei to Fouyang	Tsinan to Linchi
Chengtu to Hsichang	Kunming to Paoshan
Chengtu to Nanchwan	Taiyuan to Changshi
Taiyuan to Tatung	Taiyuan to Linfen
Wuhan to Shashih	Wuhan to Enshih
Nanning to Wuchow	Huhehot to Hailan
Sian to Yenan	Harbin to Heiho
Harbin to Ilan	

---

[1] Includes local unsurfaced roads.   [2] Includes 10,625 miles re-built.

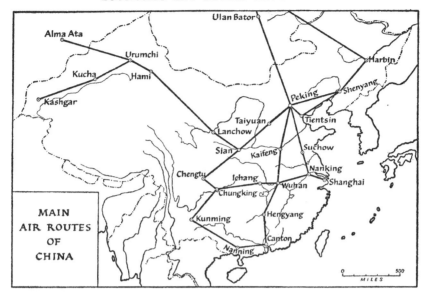

FIG. 45. Air communications

FOREIGN TRADE

Trade with countries beyond the bounds of the Middle Kingdom was never of very great significance in ancient or medieval times. Official China maintained its scorn of barbarians' attempts to supply goods to a country as well supplied as theirs. Early intercourse indirectly with Rome along the Silk Route has been described (pp. 82–86). This type of commerce, mainly through the Jade Gate over caravan routes to the west, was maintained with intermittent intensity with Persia, India and Arabia and there was some sea trade.

The visit of the Polos, and particularly Marco Polo's report of China in 1271 aroused interest in the country. The impression was given of a land of great wealth and this was undoubtedly one of the factors which stimulated the exploratory efforts of Portuguese, Spaniards, Dutch and British in the succeeding centuries. First Portugal, then Spain, followed by Great Britain and Holland, attempted to establish trading stations along the south China coast. The Portuguese, after many vicissitudes, established Macao as a trading station in 1557, paying a rental for it until 1849, when it was declared a Portuguese possession. Throughout the eighteenth century Macao was the chief port for western trade with China. The

Spaniards were able to carry on a little trade with China from Manila through Amoy. For long the British (represented by East India Company) and the Dutch were unsuccessful in their efforts to obtain recognition of trade rights in China.

In 1702 the Chinese government created a foreign trade monopoly, appointing a minister, the Hoppo, 'to be the sole broker through whom all foreigners must buy their teas and silk and must sell the few foreign products for which a demand exists' and all trade was confined to Canton. In 1720 the Hoppo was replaced by a monopoly guild or *co-hong* of thirteen Chinese merchants, through one, and only one of whom, each foreign merchant could transact business. The system lent itself readily to corruption and friction. When the Chinese government resolved to suppress the trade in opium, which was one of the most lucrative of the trade commodities, war broke out in 1840. Canton was blockaded and Amoy, Ningpo and Shanghai captured. Peace was concluded between China and Great Britain in 1842 by the Treaty of Nanking. This became the basic treaty affecting commerce between China and all foreign countries for the next hundred years, and it marks the beginning of western economic and cultural invasion of China.

The Treaty of Nanking opened five ports—Canton, Amoy, Foochow, Ningpo and Shanghai—to foreign trade, which was to be on equal footing as between China and Great Britain. There was to be a regular and fair tariff; the *co-hong* was abolished and an indemnity paid for the 20,000 chests of opium seized and destroyed. Hong Kong was ceded to Great Britain. The 'most favoured nation treatment' was accorded to the other nations (USA, France, Belgium, Norway and Sweden), who were waiting to obtain all accruing benefits. Concessions, i.e. areas of land under foreign jurisdiction, were leased to foreign powers, whose subjects also enjoyed the rights and privileges of extraterritoriality. Then a second war broke out between 1856–60, the result of which was a further extension of the above commercial rights. By the Tientsin Treaties, ten new ports, including those along the Yangtze (Nanking, Wuhu, Kiukiang and Hankow), were opened to foreign trade and shipping.

This was the setting in which the West forced open the doors of China and in which an ever increasing trade was carried on in the latter part of the nineteenth and beginning of the twentieth centuries. While trade was confined to Canton the only two articles exported to the West were silk and tea. China's imports were opium, woollen and cotton goods and small quantities of metals, perfumes, dyes and watches. This shape of trade remained even after 1842 when the south-east ports were opened, but when in 1860 the produce of the Yangtze basin was available to foreign shipping, the range of goods rapidly increased and has continued to do so. Tea remained a predominant export until the full force of Indian competition in that field was felt in the 1880's. By the turn of the century it formed only 12 per cent of the total export trade, and by 1923 only 3 per cent.

Throughout the first half of the twentieth century both volume and variety of goods continued to expand, although with marked fluctuations reflecting both the internal and international political situation. Apart from cotton goods and high quality raw cotton and cotton yarn, no single import was outstanding. Occasionally rice figured prominently. Export trade up to 1949 was confined almost exclusively to food-stuffs and raw materials. Silk remained the chief export, followed closely by soy-beans and bean-cake.

Since 1949 there has been a drastic change in foreign trade in organization, in direction and in composition. Until 1949 virtually the whole of foreign trade was in the hands of private enterprise. By 1950 70 per cent was in the hands of the state and by 1955 less than 1 per cent was being conducted by private individuals or companies. State trading was carried on officially with other states, as with India, or with trade delegations such as those sent out by Great Britain. Chinese foreign trade is now an integral and essential part of the planned economy of the state, and its flow is directed to securing and paying for those commodities immediately needed for the development of the national economy. Since the main economic objective is the building up of a modern industrial country, it is not surprising to find that the nature of that trade has changed. China still has to rely mainly on the export of foodstuffs and raw materials to secure the necessary foreign exchange to pay for imports, but recently manufactured goods (cotton and silk fabrics, light metal articles, handicrafts) have figured quite strongly, as the accompanying diagrams show.

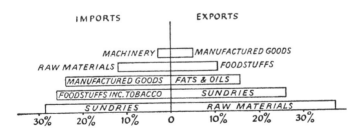

Fig. 46a. Composition of foreign trade in 1923

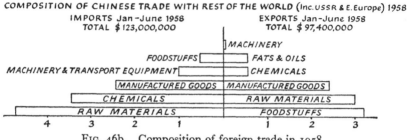

FIG. 46b. Composition of foreign trade in 1958

Grain imports, which had risen between 1946–9, ceased after that date but were resumed in large quantities following the bad harvests of 1959, 1960 and 1961. Apart from this, imports were confined almost exclusively to raw materials for industry—chemicals, manufactured goods, machinery and transport equipment.

For a full century after the Treaty of Nanking, the direction of China's foreign trade changed steadily from the old westward, landward-looking route through the Jade Gate to the new eastward, seaward orientation and to trade with Western Europe and North America. In a matter of months after 1949 this direction was once more reversed. It was through the Jade Gate to the USSR and Eastern European countries that China looked for the capital goods required for the initial steps in industrialization. Tools, machinery, tractors, oil, petroleum, chemicals, industrial raw materials and even complete factories and plant were imported from Russia. The embargo which was laid by the United Nations and especially severely by USA on a very wide range of goods during and after the Korean War served to increase this reliance on Soviet trade.

1958 *Main Imports from USSR*		1958 *Main Exports to USSR*	
million roubles		million roubles	
Complete industrial plants	664·6	Cloth	391·7
Oil products	309·5	Woven goods	348·7
Motor vehicles, equipment	246·6	Meat, butter, eggs	321·1
Iron and steel sheets	147·0	Ores, concentrates	296·0

In spite of the Korean embargo there has been a big expansion of trade with the rest of the world, particularly with Western Europe and Asia. Economic relations have been established with fifty-four countries and regions in Asia and Africa, and trade agreements concluded with eleven of them on a governmental basis. In Western Europe, West Germany's part has been particularly impressive, supplying to China fertilizers, dyestuffs, insecticides, machinery, scientific instruments and chemicals, and receiving in return egg products, feathers, soy-bean products, t'ung oil and antimony.

China's trade with countries outside the Soviet bloc[1]
(figures in millions of dollars)

	1956 Exports	1956 Imports	1957 Exports	1957 Imports	1958 Exports	1958 Imports
Canada and USA	5·6	2·5	5·0	1·5	4·0	7·2
Latin America	1·5	2·2	0·5	0·4	0·2	14·6
Middle East	20·8	26·5	18·1	42·5	19·8	34·5
West Europe	133·8	140·1	115·9	197·1	143·3	341·2
Finland	1·1	6·3	5·1	5·8	3·2	7·3
Yugoslavia	3·5	3·7	7·7	3·3	1·4	4·5
Asia	334·2	144·9	316·8	143·2	373·5	167·7
Australia and NZ	4·7	4·9	5·7	17·2	7·6	18·3
Africa	19·3	0·2	6·3	2·9	22·4	3·1
	524·5	331·3	481·1	413·9	575·4	598·4

Britain's share in this trade has grown considerably and is shown in the tables and diagram:

Sino-British Trade

Britain's Exports to China (figures in £'000s)

	1962	1963
Aircraft and parts	—	3,032
Iron and steel	718	2,899
Wool tops	854	1,864
Textiles and man-made fibres	2,361	1,274
Machinery, non-electric	512	1,120
Machinery, electric	702	550
Chemicals	694	598
Scientific instruments	128	352
Non-ferrous metals	1,328	224
Plastic materials	20	147
Metal manufactures	190	54
Road vehicles and parts	148	24
	7,503	12,290

[1] Annual Trade Review, 1949–59, *Far East Trade*, October 1959.

China's Exports to Britain (figures in £'000s)

	1962	1963
Wood, hair and other textile fibres	2,552	4,851
Bristles, hog casings, etc.	2,796	3,630
Textiles and fabrics	2,026	2,690
Non-ferrous metals[1]	1,370	1,097
Tin, unwrought	1,306	990
Hides and skins	358	987
Tea	450	653
Egg products	268	452
Chemicals	477	414
Tung oil, etc.	340	368
Essential oils	264	310
Mercury	369	199
Silver	9,872	128
Oil seeds	358	9
	22,806	16,778

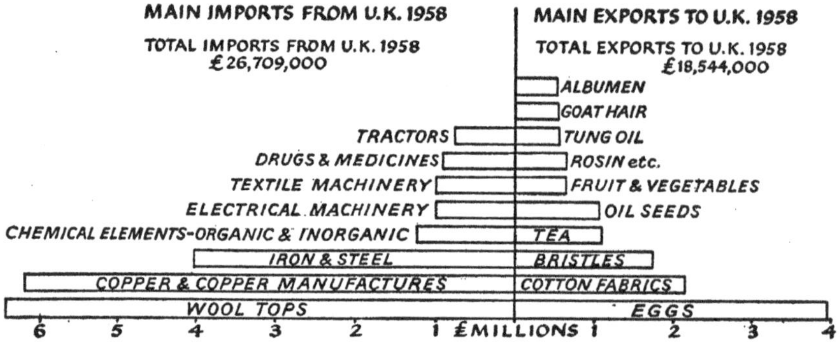

FIG. 47. Composition of trade with the United Kingdom

One way in which the Chinese government is meeting the ever-increasing demands for foreign currency is by encouraging overseas Chinese to remit payments from abroad. An Overseas Chinese Investment Company has been established, which undertakes to invest any such remittances in industrial concerns, such as sugar refineries, oil presses, etc. in South-East China, notably Fukien, which is the ancestral home of so many Chinese

[1] Excluding tin, mercury and silver.

U.K. Customs figures, quoted from the *Far Eastern Economic Review*, Hong Kong.

who have emigrated. The Investment Company also guarantees an interest of 8 per cent and undertakes not to socialize the concerns.

The government is also investing a considerable amount of capital in building up a merchant fleet. During the Japanese occupation and the civil war between 1946 and 1949, any ocean-going and river shipping that China had possessed was either sunk or fell into such a state of disrepair as to be written off. For a while the present government relied on hired shipping, but is now rapidly creating its own fleet by both purchase and by building in its own yards. China's ports—Shanghai, Tientsin, Canton, Newchwang and Dairen—are again awakening. Hong Kong, whose entre-pôt trade was so severely hit by the loss of China's trade, is again becoming a shop window for the products of the New China.

CHAPTER FOUR

Geographical Regions

For the purpose of more detailed description and analysis we have divided China into ten main regions, together with two quite small areas, Hong Kong and Macau. The two latter are not politically part of China and are therefore treated separately from the Si Kiang Basin, of which they form a part. The basis of the regional division is broadly physiographical, i.e. great river basins, plateau and inland drainage basins. Each region (except Hong Kong and Macau) is large; some are immense and are capable of almost infinite sub-division.[1]

The twelve divisions are:

1. The Hwang Ho Basin
2. The Yangtze Kiang Basin
3. The Si Kiang Basin
4. The South-east Coast Area
5. The Yunnan-Kweichow Plateau
6. The North-East (Manchuria)
7. Sinkiang
8. Inner Mongolia
9. The Tibetan Plateau
10. Taiwan
11. Hong Kong and the New Territories
12. Macau

1. THE HWANG HO BASIN

This vast river system has a catchment area of about 250,000 sq. miles and the main river a length of some 3,028 miles. It flows from west to east and drains practically the whole of north China Proper. Although the basin of the Hwai Ho is not now linked directly with the Hwang Ho, it has so often formed part of the system and its plain is for the most part continuous with that of the lower Hwang Ho, that it has been found right to treat the two together. The basin has been divided into seven major sub-divisions.

[1] Hung Fu, 'The Geographic Regions of China and their sub-divisions', *I.G.U. Proceedings*, 1952.

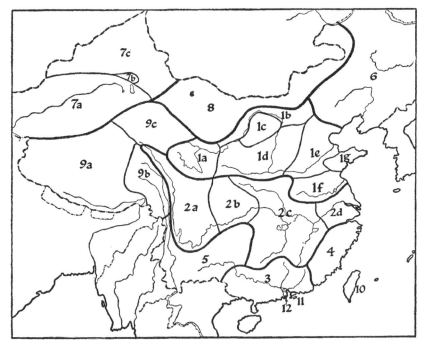

FIG. 48. Geographical regions of China

1a. *The Source*

The river rises in the Yokutsunglieh stream at a height of over 14,000 ft. in the heart of the rugged mountains of Chinghai. It is thought that in earlier geological times it drained southward to the Yangtze and that subsequent folding and faulting dammed its course and deflected it eastward in its present bed. The river in this upper tract is fast-flowing and follows a winding course through a series of fine gorges, which only recently have been properly explored and surveyed. The most magnificent of these are the Lungyang and Liuchia, both of which are now being used in river conservation works. Quite close to the head streams of the Hwang Ho and in the same mountain system is a comparatively small inland drainage basin with the Koko Nor (Chinghai) as its centre.

In the north are the snow-capped mountain ranges of the Nan Shan, which separates the steppes and the deserts of Mongolia from the Tibetan plateau and through which the river has cut to descend to the lower land of Kansu. Between these ranges are saline marshes and salt lakes. The sparse population is made up mainly of pastoral nomads.

1b. *The Mongolian Borderlands*

When the Hwang Ho issues from the Liuchia Gorge it soon changes its character from a torrential stream into a much slower and more sedate river. It first crosses the wide agricultural valleys of the Lanchow region before starting on its long northward course, known as the Great North Bend. Throughout the length of this course from Liuchia to Hokow the river marks a divide between the comparatively steep edge of the Mongolian borderlands on the north and west against which the river flows, and the Ordos and the loess regions to the south and east.

The mountainous borderlands comprise the high, rugged Ala Shan to the west of Ninghsia, the old granitic Hara (Kara) Narin Ula in the north-west, against which the Hwang Ho is turned from its northerly direction towards the east, and the Ta Ching Shan, which extend in complicated formation from the Hara (Kara) Narin Ula to the region north of Peking. The picture of the Great Wall (Plate 1) as it winds its tortuous way over mountain and valley is typical of this borderland. There is much difference of opinion as to the origin of this Mongolian edge. Wills and Hobbs maintain that the thrust causing it came from the Pacific against the Gobi block, whilst Suess, Grabau and Gregory assert that the thrust was precisely the opposite, coming from the north-west against the Ordos block. Grabau says 'The Ala Shan is an erosion remnant of a steeply overfolded series of strata, including both Sinian and Permo-Carboniferous beds . . . This overfolding is towards the east or against the Ordos platform, which formed a resistant foreland'.[1] More recent research favours this latter theory.

The river all the way from the Chingtung gorge below Lanchow to Hokow is wide, slow-flowing and very shallow. The average gradient for this reach of the river is 1 in 5,000. It is navigable only for craft such as rafts or 'boats' of inflated skins, drawing less than three feet of water. These are floated downstream as far as Hokow where they are broken up, the timber being sold and the skins returned by pack animals or by light craft, laboriously towed upstream by trackers. The whole process is very reminiscent of transport down the Tigris in Babylonian times.

The river's régime is very seasonal and is also very variable. Rainfall is sparse and there is a very marked summer maximum. Lanchow has a mean annual rainfall of 16·7 in., 9·9 in. of which falls in July and August. Moreover, torrential downpours and cloudbursts are characteristic. As much as 5 or 6 in. may fall in a single day with disastrous results to the crops. No tributary of note joins the river in this section.

[1] A. W. Grabau, *The Stratigraphy of China* (Peking 1925), Part II, p. 305.

Down the river below Chingtung is the rich and densely populated agricultural region of Ninghsia. This is an elongated oasis, stretching about 60 miles along the river where the first irrigation works were carried out in the Ching dynasty (221 BC). These works have been maintained in a fluctuating state of repair ever since, and now form the basis of big new developments. Wherever water is available, heavy crops can be grown during the five months growing season. Rice, wheat and millet are the main cereals. Ninghsia is also renowned for its temperate fruits (apricots and peaches) and *kwa* (melons). The Ala Shan to the west, which were once well covered, have been seriously deforested.

It is here in Kansu that the greatest concentration of Chinese Moslems (*Hui Chiao*) is found. They number some 3,550,000 according to the 1953 census and are of Turki extraction. They have constituted one of the most turbulent corners of the country, their last big rebellion being in 1894–5, when about a quarter of a million people were slain.

There has been some attempt at irrigation and development where the Hwang-ho turns eastward beneath the Kara Narin Ula in the Hotao district. This was the work partly of Belgian Roman Catholic missionaries and partly of the China International Famine Relief Commission in 1929, but owing to aridity and to political unrest, the area has not prospered greatly.

At the eastern end of the Great North Bend of the river, lying between the cities of Saratsi, Huhehot (Kweihwa) and Hokow, there is a triangle of the most fertile land outside the Great Wall. Most of the cultivation was without the aid of irrigation until the CIFRC built considerable works in 1929 in an endeavour to avoid future famine disasters in this area. Saratsi's average annual rainfall is only 13·5 in., 10·6 in. of which falls in the four summer months from June to September, but this, too, is a region of great variability. The extension of irrigation works is now going on energetically and has reached as far west as Paotow. The main crops are millet, kaoliang, spring wheat and hemp, while rice is increasingly grown under irrigation.

Paotow, which until very recently was an old market town, has suddenly flourished into a great industrial centre. The Ta Ching Shan to the north are largely of Carboniferous and Jurassic sandstones and shales, and are very rich in anthracite and bituminous coal. There are also fair deposits of iron. Paotow was therefore selected by the planners as one of the sites for the development of integrated plant for the manufacture of heavy iron and steel products. Already it is producing more than 1 million tons of steel per annum. The Peking-Paotow railway now runs the length of the Mongolian borderland through Ningsia to Lanchow, linking up with the Lunghai line.

Huhehot (Kweihwa) is an old city, typical of so many in that it has a walled Manchu garrison city quite separate from that of the *peh hsing* (people). During the last decade or two the wall of many cities has been demolished. Huhehot is now the centre of the University of Inner Mongolia and a big new city is developing.

1c. *The Ordos*

Bounded by the Hwang Ho in the west and north in its Great North Bend, and on the east and south by the remnants of the Great Wall, is the Ordos, the most arid part of China Proper. In no part of it is the average annual rainfall more than 10 in. and in some parts it is considerably less. What rain there is often comes as torrential downpours, turning the dry gullies into short-lived turbulent streams. Consequently there are no tributaries worthy of note joining the Hwang Ho from this desert and semi-desert region. The whole area is studded with salt lakes and glistening saline pans. The soil is mainly of immature calcium greyearths, the fine dust particles of which are carried in a general direction south-eastward, although the persistently strong winds are remarkably unpredictable in direction. George B. Cressey maintains that the loess of the region immediately to the south-east is derived from the Ordos and not from the Gobi farther west.[1] He points out in support of this theory that there are no loess deposits in Northern Shansi and Northern Hopei on to which the Gobi abuts. The heavier material, which cannot be air-borne, remains as sand dunes, particularly in the north-west and south-east. 'Due to the slow but steady migration of sand-dunes from Ala Shan, the southern area covered by steppe is decreasing in extent.'[2]

The natural vegetation is scanty steppe bunch grass with a few scattered tsaidams, areas where the grass cover is more abundant and continuous. Population, which is very sparse, is Mongol and nomadic, moving with their herds of sheep, goats, horses and some cattle over the poor pasture. The new government has not yet tackled this region, except along its perimeter and therefore these pastoralists continue to live the life of typical Mongolian nomads, having yurts for dwellings, sheepskin as their main clothing, and milk, butter, cheese, mutton and brick tea (from Hankow) as their staple food.

[1] G. B. Cressey, 'The Ordos Desert of Inner Mongolia', *Dennison University Bulletin*, Vol. 33, No. 8.

[2] Tsui Yu-wen, 'Problems of Soil-preserving Vegetation in the Middle Reaches of the Hwang Ho', *Academia Sinica*, 1957.

The Great Wall, which divides the Ordos from the loess in a quite graphic manner, has also laid down a rough line of demarcation between the Mongolian pastoral and the Chinese arable way of life. There are a few oases within the Ordos into which the Han Jen of China have pushed their precarious way, and have managed to establish themselves. Arable farming is continually confronted in this arid area with the danger of alkalinity. Salts are brought to the surface by capillary attraction and are not dispersed if flushing is insufficient.

1d. *The Loess*

Bounded by the Ordos and the Great Wall in the north and west, by the Tsinling Shan in the south and the Tai Hang edge in the east and covering the whole of the province of Shensi, most of east Kansu, west Shansi and west Honan is one of the most remarkable regions of the world. Here an aeolian dust of minute yellowish-grey grains has been laid down in a thickness varying from a few feet to over 250 feet. These deposits of loess, which are unstratified, have masked and transformed the former old, dissected peneplain. Wide valleys have been filled and converted into broad plains through which the former peaks and ranges emerge, rather like nunataks.

As we have seen, Cressey contends that this air-borne dust has come from the Ordos. Buxton, however, points out that there are wide areas of the Gobi today which are devoid of sand, and that all light loose material has been carried by the wind and laid down in the more humid areas to the south-east.

Lowdermilk's micrometer analysis[1] of the loess, given below, reveals the fineness of the material, which in most instances is so smooth as to be capable of being rubbed into the pores of the skin like talcum powder. The fact that the grains are angular, points to their being aeolian and not sedimentary in origin.

[1] Quoted from D. Buxton, *China, the Land and the People* (Oxford 1929), p. 268.

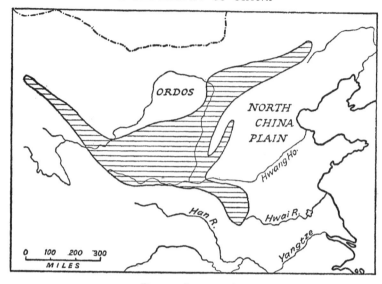

FIG. 49. Loess region

Class	Average Diameter of Particle Sample 1	Sample 2	% Sample 1	% Sample 2
Fine Sand	0·13 mm.	0·10 mm.	1·59	1·00
Very fine sand	0·065 mm.	0·063 mm.	27·44	25·00
Silt	0·033 mm.	0·030 mm.	50·97	54·00
Clay	0·0035 mm.	0·004 mm.	20·00	20·00

An outstanding characteristic of the loess is its proneness to vertical cleavage, which results in the precipitous valleys and unique cliff landscape so typical of the area. Several theories have been advanced to explain the causation of this cleavage. The most probable of these is that at the time of deposition the natural vegetation of the region was abundant steppe grass, which was successively buried and fresh grass established. The old buried grass decayed, leaving minute vertical hollow tubes along which lines of cleavage occur. A further characteristic is that on exposure to the air the vertical cliff face tends to form a thin cement-like covering.

This nature of the loess has been utilized by the people to solve their housing problems, because it is an easily worked material. The cliff face is cut into to form houses, sometimes of two and three storeys. While not to be recommended for their ventilation, they provide warm and dry quarters in the winter and are remarkably cool in summer. Their great defect lies in the fact that this is an area subject to earthquakes, and these cave dwellings are terribly vulnerable. An earthquake brings the cliff faces down in great landslides, burying all beneath them and destroying the fields which often lie above. In the 1920 earthquake about one quarter of a million people perished either directly or through the subsequent famine.

The climate of the region is one of extremes. Winters are bitterly cold, most areas receiving the full force of the cutting, dust-laden winds from Mongolia. An old Chinese description of the district runs: 'The sky is grey; the wilderness misty. The grass bows low to the wind where sheep and cattle are grazing' but the skies can also be clear and bright in winter. Although they coincide with the rains, the summers are blisteringly hot. Although the region is not as dry as the Ordos and the regions farther north-west, rainfall is still sparse and the inshore monsoon only too prone to failure. Consequent on this and also on the fact that this region is remote and lacking in communications, it is a notorious famine area. Andersson writes of it thus:

'The most fatally incalculable factor of the seasons is the rainfall. Since reckless felling has destroyed the last remnants of the primeval forests, which, by evidence of the Stone Age deposits, once covered the land, the treeless loess plain has become exceptionally sensitive to changes in rainfall. If the normal light rainfall fails, there is no reserve of moisture in the plateau, which is drained by innumerable ravines. If, on the other hand, the summer rains come with the violence of a cloudburst, as not infrequently happens, the ravines are widened with catastrophic rapidity. New miniature ravines are formed in a single night of rain, houses are threatened and roads diverted. Most feared is drought, which is synonymous with famine.'[1]

	January	July	Rainfall
Taiyuan	17°F.	77°F.	13·8 in.
(2,592 ft.)	−8·4°C.	25·0°C.	11·6 in. (June to Sept.)
Sian	33°F.	86°F.	20 in.
(1,095 ft.)	0·6°C.	30·0°C.	13·1 in. (May to Aug.)

The natural vegetation is bunch steppe grass with some steppe forest on the higher mountain slopes. There are some herbaceous plants in the wetter parts—bramble, smart weed and wild chrysanthemum and some

---

[1] J. G. Andersson, *Children of the Yellow Earth* (London 1934), p. 169.

wild clove. In the south-east and in the Wei valley there are remnants of broadleaf and evergreen forest, but the whole region suffers from deforestation and stripping of soil cover. The result has been soil erosion of a very serious character, especially in the north and west. The light, uncoordinated loess, when bereft of its cover, breaks down with alarming ease and disastrous erosion follows. A local saying sums it up: 'Poor hills; furious waters'. Deep ravines and gullies are carved, carrying away the farmers' fields above and burying the fields in the lower valleys.

Apart from the valley bottoms there is little irrigation at present. The hillsides have been terraced, but not adequately, and they are very vulnerable to both gully and sheet erosion. Much attention has been given recently to better farming to meet the danger: stronger terracing, contour ploughing and intercropping. It is in this region that the big efforts at establishing shelter belts mentioned above (p. 180) are being made.

These deep ravines make communications in the region very difficult. They cut deeply into the flat plain and render transverse movement in many places virtually impossible. Loess is so soft that constant traffic along unsurfaced roads or tracks has cut them down so that they are sometimes 10 or even 20 ft. below the surface of the surrounding countryside. The construction of properly engineered and surfaced roads in this area will prove a boon.

Below this great blanket of loess lie the great coalfields of Shensi, Shansi and Honan, whose combined reserves stand at over 40,000 million tons of anthracite and over 160,000 million tons of bituminous coal. Until recently very little of this wealth had been developed owing to its remoteness and lack of communication. Apart from fairly extensive mining and steel works around Taiyuan and Tatung, exploitation of this coal is confined to the periphery. There is little doubt that there will be widespread development here in the near future.

The Hwang Ho, which is a wide, shallow, slow-flowing river from Chungwei to Hokow, turns southward at that point and enters the loess. Its gradient is greatly increased and it becomes fast-flowing. It cuts a deep, narrow valley into the loess and is unnavigable from Hokow to Lungmen. It is in this compartively short sector that the Hwang Ho receives nearly 50 per cent of its silt load. Its main left bank tributary, the Fen, rises in the sacred Wutai Shan of east Shansi and, following the general NE-SW trend of the mountains, flows in its upper and middle reaches in a narrow graben or faulted trough. There the provincial capital, Taiyuan, stands. The lower valley of the Fen is wide, undulating and fertile, being filled with loess.

The Wei Ho joins the Hwang Ho as its main right bank tributary. It flows from west to east along the foot of the steep northern face of the Tsinling Shan and is itself joined by two left bank tributaries, the Ching and the Lo, which descend from the gentler slopes of the Shensi plateau to the north. All these three rivers carry very heavy loads of silt down to the main stream. The Wei valley is broad and generally flat or undulating. Both the Wei and the lower Fen valleys have been irrigated from early historical times and formed the earliest granary of Classical China. The irrigation works of the Fen have not been maintained in as good repair as those of the Wei, but both valleys produce heavy crops of winter wheat and millet as well as good quality cotton. Wei valley fruit—walnuts, persimmons, peaches, plums, apples and pears—is renowned throughout the country.

In the heart of the Wei valley and standing on the foothills of the Tsinling to the south of the river, is the ancient imperial capital city of Ch'ang-an. The present provincial capital of Sian stands close to the old site. It served the Chou dynasty from 1122–255 BC, when it was known as Haoking and it was retained as imperial capital by Ch'ing Shih Hwang Ti(221–206 BC) and by the emperors of the Former Han (206 BC–AD 25). Ch'ang-an lost its imperial status for nearly six hundred years but regained it under the Sui (AD 581–619) and T'ang (619–907) rulers. Thereafter it has never been more than a provincial capital.

For more than 100 miles after its confluence with the Wei Ho, the Hwang Ho is a difficult and dangerous river for navigation. Its gorges between Tungkwan and Sanmen do not compare in grandeur with those of the Yangtze but they have been enough of an obstacle to navigation and a barrier to movement to give a sense of security to rulers falling back on Sian (Ch'ang-an) when pressed from the east. Here the river cuts its way between the Tsinling Shan on the south and the Chung-tiao Shan, which form the southern edge of the Shansi plateau, and descends at Sanmen to issue on to the North China Plain.

1e. *The North China Plain*

In fairly recent geological times the river at Sanmen issued, not on to the vast flat plain which now exists, but into a sea, an extension of the present Yellow Sea and Pohai, which then lapped against the Tai Hang edge of the Shansi plateau to the north and the Hsung (Sung) Erh Shan and Fu Niu Shan to the south. Shantung stood as a group of islands in the midst of this sea.

Throughout the millennia which followed, the Hwang Ho has continuously poured out its load of silt, marine sands have been laid down and loess from the west has been carried by the Mongolian winds, combining to build what is now the North China Plain. The mind boggles at the amount of material that has been deposited, for this is a trough, a geosyncline, into whose maw seemingly endless food can be poured, since the rate of sinking maintains an equilibrium with the rate of deposition. We have seen that the Hwang Ho's silts and sands are at least 2,800 ft. thick. Some idea of the speed and extent at which the Pohai is being filled can be gained from the recorded fact that between 1949 and 1951 the coast at the mouth of the Hwang Ho was pushed 6 miles seaward in a fan having a 25 mile front. The sea is so shallow and the shoreline south of Taku so indeterminate that it is quite difficult to detect sea from land. Fishermen on stilts can be seen fishing miles out to sea. In the north the Five Rivers, chief of which is the Hun Ho (Muddy River), which flows past Peking, all make their contribution of fill to the Pohai.

The North China Plain has quite distinct borders on the north and west. On the north are the mountain ranges of the Peking Grid, the Heng Shan and the mountains of northern Hopei (former Jehol) all of which rose steeply from the plain. Two gates through this northern mountain wall have been historically important. The former road of entry of Tartar hordes was through the Nankow Pass and was one of the main reasons for choosing Peking as the site of the Imperial capital. It now forms the line of communication by road and rail, via Changkiakow (Kalgan) to Inner and Outer Mongolia. The Jehol Mountains descend nearly to the sea at Shanhaikwan, leaving only a narrow corridor of lowland, which forms the main line of communication between China and the North-East. It was through this gateway that the Manchus came to conquer China and to establish their dynasty in the seventeenth century.

The western border of the plain is equally well defined by the Wu-tai Shan and the Tai Hang Shan north of the river, and the Sung Erh Shan south of the river. The Shantung hills and the sea coast mark the eastern limits of the plain, but on the south no clear-cut divide from the basin of the Hwai is discernible. The lower reaches of the Hwai have so often been part of the Hwang Ho when the latter has flowed to the sea south of the Shantung peninsula that we have seen fit to include it in the Hwang Ho basin, although, if the conservancy schemes for both rivers, which are now in hand, prove successful, the two rivers will go their separate ways in future.

Even though this great flat plain abuts the sea, its climate is one of extremes, but less severe than in regions farther west. The plain has some shelter from the Mongolian north-west gales by the Tai Hang Shan; nevertheless the cold off-shore winds from the Mongolian high are the winter arbiters. Winter days are usually cold, clear and dry, but the northern part of the plain in particular is also subject to severe dust storms, when soil cover from the 'dust bowl' to the west is carried out to sea. Little describes these: 'I have, myself, travelled in a steamer compelled by a storm of impalpable dust to anchor in the Inland Sea of Japan, 500 miles distant from the coast of China, as though fog-bound.'[1] Snowfall is usually light. A heavy fall is a matter for rejoicing by the farmers as it provides a protective cover for the winter crops and is regarded as an assurance of good harvests to come.

Summers are hot and rather humid. Annual rainfall is between 20 and 25 in. with a very marked summer maximum, but once again with the disadvantage that it may fall in torrential downpours which do far more harm than good. Although the rains are not as variable as farther west, they are nevertheless unreliable, and the June rains are liable to fail. Then everything is burnt up; the winter wheat is scorched and salvation lies only in the late rains of July and in crops of millet and *kaoliang*. The southern part of the North China Plain is less subject to temperature extremes: it enjoys a greater rainfall with a better distribution and a greater reliability than the north.

	January	July	Average annual rainfall
Peking	23°F.	76°F.	24·9 in.
(131 ft.)	−5·1°C.	24·5°C.	18·7 (June, July and Aug.)
Chefoo	23°F.	78°F.	24·6 in.
(10 ft.)	−5·1°C.	25·6°C.	17·6 (June to Sept.)
Kaifeng	30°F.	82°F.	46·6 in.
(328 ft.)	−1·1°C.	27·6°C.	32·1 (April to Sept.)

The soils of the plain are generally pedocals of unleached calcareous alluvium. The more fertile areas are those fringing the higher land of the Tai Hang Shan, Fu Niu Shan and Shantung, which are better drained and are the sites of early civilization on the plain. The less well drained regions nearer the coast are often marshy and saline.

After it emerges from its gorges between Tungkwan and Sanmen and reaches the plain, the Hwang Ho becomes a slow-flowing, meandering river, which fills only a small part of its wide, shallow bed during the dry

[1] A. Little, *The Far East* (Oxford 1905), p. 29.

winter, but with the summer rains it is transformed into a fast-moving leviathan, full of menace. Some idea of this change in régime can be gleaned from the figures of discharge, which record a maximum flow of 25,000 cu. metres per sec. (883,000 cu. ft. p. s.) and a minimum of only 245 cu. metres p.s. (8,650 cu. ft. p.s.).[1] Such variation in volume between summer and winter has led to constant inundation, which in fact has built up the plain over which it flows. Chinese history records more than 1,500 inundations in the last 3,000 years and twenty-six changes in course, nine of which were of a major nature. Up to 602 BC the Hwang Ho flowed north and found an outlet near to present Tientsin. It then made a drastic change and flowed out to the Yellow Sea south of the Shantung Peninsula and there it remained until AD 70 when it took up a course much along its present bed. Between AD 1048 and 1324 the Hwang Ho moved back to its earlier lines and flowed out again near Tientsin, after which it again found its outlet south of Shantung. In 1851 the Hwang Ho turned north and flowed into the Pohai. In 1938, when the Japanese invaded the heart of China, the Kuomintang Government deliberately cut the southern banks of the river in the hope of checking the enemy advance. About 54,000 sq. kilometres (20,744 sq. miles) of land in the northern Hwai basin were flooded, resulting in the death of nearly 900,000 people. The river was thus turned south of Shantung once again. It was returned to its former northern course in 1947 as a result of United Nations action.

Endeavours to control this unruly river have been an important function of Chinese government all through the centuries. Unhappily Li Ping's advice, 'Keep low the dykes, keep deep the channels' has not been followed here, and reliance has been placed exclusively on dyke building. These dykes have not been built as one concerted scheme but piecemeal, locality by locality. Generally their constructors have favoured building the dykes 5 to 8 miles (8 to 12 km.) apart, thus allowing the river plenty of room with the idea that when it was in spate it would be accommodated. The trouble has been that, even if this has been achieved without the dykes breaking, the river is very slow and consequently deposits much of its load of silt on the bed. As time has gone on the bed has been continually raised in this way and this, in its turn, has necessitated constant raising of the dykes, with a result that today the bed of the river stands higher than the surrounding country. As one approaches the river, it looks like an endless, uniform range of low hills. The dangers of such a situation are not

[1] *Flood Control Journal No. 1* (September 1949), United Nations Economic Commission for Asia and the Far East.

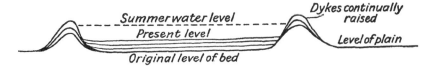

hard to see. If, through laziness on the part of the local farming community or through political unrest, the dykes are allowed to fall into disrepair or there is an exceptional summer spate and a breach is made, then the ensuing flood is doubly serious. The country around is inundated to a greater or lesser degree and the season's crops destroyed, but worse, when the spate is over and the level of the river falls, the flood waters cannot return to the old bed. They have to find a new line of drainage; and hence the river may make one of its periodic changes of course. It is significant that, because of the raised bed, the Hwang Ho receives no tributaries below Kaifeng, the point below which most changes of course have taken place. The cost in human loss of life and suffering through the centuries has been enormous. The cost in material loss each year—loss of harvest, ruined fields, and so on—is staggering. Flood spells famine and disease. The river richly deserves the stigma attaching to its name 'China's Sorrow'.

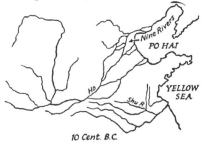

This building up of the river's bed above the surrounding plain can better be appreciated when the summer load of the river is examined. It is no exaggeration to say that it leaves the gorges at Sanmen like a thick, yellow soup. The Chinese have a saying that 'If you fall into the Hwang Ho you never get clean again'. The average silt concentration per cubic metre is 34 kilograms (2·26 lb. per cu. ft.) as

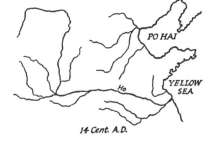

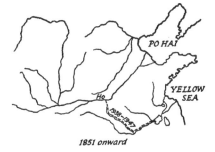

FIG. 50. Some changes in the course of the Hwang Ho

compared with 10 kg. (0·66 lb. per cu. ft.) in the Colorado, 4 kg. (0·26 lb. per cu. ft.) in the Amu Darya and 1 kg. (0·065 lb. per cu. ft.) in the Nile.[1] It is estimated that the load carried annually is 415 billion cu. metres and that 40 per cent of this is deposited in the river bed, while 60 per cent reaches the seaboard. Nearly 90 per cent of this silt comes from the loess region and enters the main river between Hokow and Shanhsien, figures which give some idea of the seriousness of soil erosion in Shansi and Shensi, where an average of 3,700 tons of soil is carried away annually from every square km. of land. This is 27 times greater than the world average. The reaches above Hokow contribute only 10·9 per cent of the total load.

To meet this eternal problem the government has evolved a scheme, the general lines of which were made public in 1955.[2] The Plan, which in many ways is similar to the Tennessee Valley Authority scheme, has four objectives: flood control, the generation of hydro-electric power, the provision of vast irrigation and the conversion of the Hwang Ho into a navigable river. For purposes of conservation, the river has been divided into three reaches; upper, middle and lower, and the main attention has been concentrated on the middle reach from Lungyang Gorge, above Kweiteh to Taohwayu below Sanmen.

The planners place their faith mainly in the construction of silt-retaining dams to control flow. To this end the 'Staircase Plan' has been designed, which envisages building a series of forty-four dams on the main river in this middle section, the lower part of which we have seen is the main contributor of the silt load. Two other dams, mainly for irrigation, are to be constructed below Taohwayu. In addition, many smaller dams are to be built on tributaries. It is estimated that the whole scheme will take seventy years to complete. All the dams have silt retention as their primary purpose. There are to be four big multiple-purpose dams, three of which are already completed or nearing completion. Two are in the Chinghai mountains. Lungyang dam is to supply power to the Tsaidam; the Liuchia dam, which creates a reservoir of 4,900 million cubic metres gives power and irrigation to Lanchow. The really spectacular achievement, however, is the building of a huge dam at Sanmen. The name Sanmen means 'Three Doors'. Here there are two rocky islets dividing the river into three streams, which have been named. 'The Gate of Man', 'The Gate of the Gods' and 'The Gate of the Ghosts'. This spot has been chosen for the building of a dam, which is 295 ft. (90 metres) high and which will retain a reservoir or

[1] Teng Tse-hui, *Report on the Multiple Purpose Plan for permanently controlling the Yellow River*, 18th July 1955.

[2] ibid.

FIG. 51. The Hwang Ho and Hwai, showing gorges, dams and water conservancy

lake of some 907 sq. miles (2,350 sq. km.)—greater in area by 150 km. than Taihu in Kiangsu—and will hold 36,000 million cubic metres of water. The lake will reach back along the river course beyond Tungkwan and will convert that torrent stretch into a navigable waterway. It will thus be capable of reducing the biggest recorded flow of 37,000 cubic metres per second to 8,000.[1] It is estimated that, given present conditions, the lake will be silted up in 50 to 70 years, but if other soil conservation schemes are successful, this time will be greatly extended.

When the full scheme of 46 dams is operating, 23 million kW. will be generated, giving an average annual output of 110,000 million kW. The Sanmen and Liuchia stations will produce annually 9,800 million kW. and 6,600 million kW. respectively in the initial period.

By means of these high dams and the huge reservoirs which are thus created, the planners reckon to extend the present irrigated area eventually from 16,500,000 *mow* (2,750,000 acres) to 116 million *mow* (19,330,000

[1] The completion of the Sanmen scheme has been retarded by the withdrawal of USSR technical experts.

acres). The head of water in the lakes will remain high enough to provide irrigation water throughout the year.

Finally the scheme envisages a river which will be navigable for 500 ton tugs from its mouth to Lanchow during the whole year. If this is achieved, it will enhance the economic wealth of the whole basin enormously.

This grandiose multiple-purpose plan is being supported by the far-reaching afforestation campaign described above (p. 179) and the drive for better, more careful cultivation by means of better terracing, contour ploughing and contour cropping, and giving up the cultivation of steeper slopes and their return to trees and grass, long advocated by Lossing Buck.[1]

This, very briefly, is the far-reaching, long term plan for the control of the Hwang Ho. To meet the immediate dangers and needs, much work has been done on repairing and strengthening the existing dykes. Where possible stone facing has replaced the kaoliang stalks previously used. A systematic war has been waged on burrowing animals, especially foxes, which abound in the north. A diversion channel from the Hwang Ho to the Wei (Hopei), to carry some of the former's flood water was completed in 1953, and a large retention lake was formed between the two rivers.

Belonging to the North China Plain but not part of the Hwang Ho drainage basin are the Five Rivers of the north, which combine to flow into the Po-hai at Ta-ku, the port of Tientsin. The chief of these rivers is the Yungting, which flows past Peking and which has been the cause of much flooding on the alluvial plain. A dam, 150 ft. high and 950 ft. long, has been built at Kwantung above Peking. This has the four-fold purpose of flood prevention, the provision of irrigation water to the plain, the generation of power and the provision of a water supply to Peking.

Any project involving the formation of large reservoirs and the inundation of large tracks of land on the North China Plain will also involve problems of human resettlement, for here is the greatest congregation of farming communities in the world. The North China Plain, which is almost identical in extent with Buck's winter wheat-kaoliang area, covers about 150,000 sq. miles (388,500 sq. km.) of which some 70 per cent or 105,000 sq. miles are cultivated. The population of the plain is over 100 million, of which approximately 80 per cent are engaged in agriculture. In many rural districts there are over 1,000 persons per square mile of cultivated land but considerably less in the saline areas near the coast. The population is very widely dispersed throughout the plain in hamlets and villages. Here and there are larger market towns, usually walled. These walls are

[1] Lossing Buck, *Land Utilization in China* (Shanghai 1937).

now largely being pulled down and the material used for road making and building. The people are almost entirely Han Jen with a few Manchus, now indistinguishable from the rest, scattered amongst them. The northerner is a taller, slower, more stolid person than his T'ang countryman in the south.

Even before 1949 this was a region of larger fields and larger farms than farther south, although not so large as those of the loess region. The average size of fields in the North China Plain was 1·26 acres as compared with 0·17 acres in the rice-tea area south of the Yangtze and 0·12 acres in the double cropping rice area of Kwangtung. Farms, too, were larger, averaging 5·1 acres and were very much larger than this in the poorly drained and saline areas. Farm ownership here was much more general than in the south. Standards of production and of living were generally higher here than in other areas north of the rice growing line. Although the basis of production has changed from individual ownership and enterprise to communes and a socialist economy, and although in most villages communal kitchens and schools, etc. have been built, the general appearance of the villages has not changed appreciably. Villagers continue to live in the same simple, mud-brick houses with their brick stove beds and their walled courts or yards.

The main crops cultivated are winter wheat, millet, kaoliang, barley, soy-bean, corn, sweet potatoes, peanuts and a large variety of vegetables, including cabbage, turnip, onions, garlic, radishes, cucumbers, spinach, string beans, peas, melons and squashes. Until recently about 39 per cent of the cultivated land was double cropped. The main summer crops are sweet potatoes, peanuts and soy-beans, which occupy about 60 per cent of the land, millet 30 per cent and vegetables 10 per cent.[1] Winter wheat and barley are grown from November to June, or the land may be left fallow through lack of fertilizer since farms are lightly stocked and farmyard manure is scarce. Generally speaking, before 1949 the poorer the family the more sweet potatoes were grown and the less wheat.

Wheat is a good cash crop but is harder to grow and requires more cultivation and more manure than sweet potatoes. The land is ploughed and fed in late August and September as soon as the summer crop of soy-beans, millet or sweet potatoes has been harvested. By the end of November the shoots are strong. If at this time there is a heavy snowfall there is great rejoicing, especially at Chinese New Year, for the snow provides a protective cover for the crop during the bitter cold of January and Feb-

[1] Martin C. Yang, *A Chinese Village: Taitou* (London 1947).

ruary. Harvesting is done at about the June festival of Tuan Wu (Double Fifth), i.e. 5th day of the 5th month.

Foxtail millet and kaoliang are the main summer cereals and together with sweet potatoes they provide the staple food of the country folk. All three crop heavily, and as all three stand up to drought conditions well, they give some security against the failure of the winter wheat. Peanuts, which are an alternative crop to sweet potatoes, are sown in May. They need a sandy soil and dry weather. Harvesting takes place in October when the plant is deeply hoed, and the vines are lifted as a whole with the shells still clinging.

Marketing facilities on the North China Plain have been better than in other parts of China. The dirt-mud roads or tracks over the flat land have made communication easier than in the hilly or mountainous west and south. Moreover, railway communications had their earliest development over the Plain.

While the land and climate, the hardworking, tough peasantry and, to a large extent also, their crops have remained the same, big changes have taken place in organization and farming technique since 1949. The flat or undulating plain lends itself readily to the kind of amalgamation that has taken place under the co-operatives, collectives and communes. Fields, the average size of which was little more than an acre, are now often fifty acres and more. This increase in the size of fields makes possible the economic use of farm machinery. It is here on the North China Plain that mechanization is being given priority.

Irrigation, which was practised on not more than 10 per cent of the Plain, has now been greatly extended largely by sinking thousands of wells and by the construction of local earth dams. The emphasis which is now laid on the greater use of chemical fertilizer is resulting in much less land being left fallow. The problem of graves has been tackled vigorously in the north. Lossing Buck estimated that at least 2 per cent of good arable land was devoted to ancestral graves. Often these graves were in the middle of the field, which was not only a waste of valuable land occupied by the grave itself but also rendered ploughing much more difficult. Now most of the graves have been removed to cemeteries on waste land. Another change is the greater use of women's labour in the fields. In many places on the North China Plain formerly it was customary that women's work in the fields should not begin until threshing of the winter wheat or barley. Since the formation of the co-operatives and then the communes, women have been working continuously throughout the year alongside the men.

The North China Plain is rich in imperial capitals. It includes within its borders the most ancient Chinese capital and three cities that, in their time, have been imperial capitals. The Great City Yin, capital of the Shangs, near Anyang, stood on the Hwang Ho when it flowed north, close to the Tai Hang edge, and found an outlet in the 'Nine Rivers' near present-day Tientsin. Loyang, now capital of the province of Honan, stands at the eastern end of the east-west gate of Tungkwan and was the imperial city of the Later Han emperors (AD 25–196). Today it has developed into an important industrial centre, specializing in car and tractor manufacture. Between AD 960–1127 Kaifeng was the imperial capital of the Northern Sung and was then called Pienking. We have seen that it was sited at the point from which most of the changes of course of the Hwang Ho have occurred. In consequence it is a point of some strategical importance. Movement along a north-south line east of Kaifeng is more difficult than to the west and therefore there is something of a bottle-neck between Kaifeng and the Fu-Niu Shan.

The third imperial seat of the plain, Peking, is a comparatively new city. It was first built by the Liao emperors (AD 937–1123) and named Yenching. When the Kin Tartars (Golden Horde) ousted the Liaos they adopted Yenching and renamed it Chung-tu. However, Peking did not attain true imperial status until the Kin in their turn were driven out by the Mongols in AD 1234 and the Yuan dynasty was founded. Kublai Khan began to build the city of Cambaluc, which means the Khan's City, near the old site. With a vast empire stretching right across Asia to Europe and the whole of China to the south, the importance of Peking as the governmental centre is abundantly clear. All subsequent governments, Ming and Ching, Republic and People's Republic, have recognized this, and Peking has retained its status ever since, except for one short break between 1927 and 1949.

Peking is now the hub of a vast governmental and administrative machine whose controls reach out to all corners of the realm—to Tibet, Chinghai, Sinkiang, Heilungkiang and to farthest Yunnan—with an intimacy and effectiveness never known before in China. Since its foundation it has always been one of the chief centres of education. Christian missions in the first half of this century played a prominent part in the promotion of western learning here by the creation of the PUMC (Peking Union Medical College) and Yenching University, which is now the campus of the National University of Peking. Since 1949 Peking has been the site of feverish new development in higher education, especially on the technical side. No less than 17 colleges and technical institutes, usually catering for about 5,000 students each, have sprung up in the city's environs.

The people of Peking have regained an acute consciousness of, and pride in, their historical heritage. All the architectural gems which have escaped the ravages and neglect of the last 40–50 years—notably the Forbidden City, the Summer and Winter Palaces, and the Temple and Altar of Heaven—have been renovated and are now museums, open to the public, which throngs to see them. Like most modern capital cities, Peking is rapidly developing as an industrial centre, having coalfields near at hand and now hydro-electric power from Kuanting.

Tientsin, which served as port for Peking and is now itself served by Ta-ku, is developing into a big industrial as well as a commercial centre.

1f. *The Hwai Basin*

The Hwai River, which is about 680 miles long, flows over a vast alluvial, lake-studded plain and drains 67,200 sq. miles, more than two thirds of which is under cultivation. There is no appreciable watershed between the Hwai and Hwang Ho on the north nor between the Hwai and the Yangtze in its lower delta reaches. There are, however, distinct differences between the Hwai and the Hwang Ho. We have seen that the latter receives no tributaries in its course across the plain. The Hwai receives a large number of long left-bank tributaries, rising in the Fu-niu Shan, all of which follow roughly a NW-SE direction and a lesser number of shorter right-bank tributaries in its upper reaches from the Ta Pieh Shan. The tributaries and the main river all drain into the Hungtze Hu, whence the waters have often found an indeterminate route to the Yellow Sea, and have even on occasion flowed south into the Yangtze. The Hwai is not so burdened with silt as the Hwang Ho since the Fu-niu Shan and Ta Pieh Shan from which it derives most of its water have comparatively little loess covering.

The Hwai Basin being farther south, the climate is more equable than that of the North China Plain: winters are shorter, the growing period longer and the summer rains heavier. Nevertheless, the crops grown are more closely akin to those of the northern plain than of the Yangtze Basin. Winter wheat, millet and kaoliang are more characteristic than rice, and for this reason, if for no other, the Hwai and the lower Hwang Ho basins should be classed as one region.

The fact that two-thirds of the land is under crops proclaims how fertile a plain this is, but like the North China Plain, it is very subject to flood. The Hwai has not built up its bed above the surrounding level of the land as the Hwang Ho has done, yet most of the country is so low-lying that great stretches of the river have been dyked. In consequence, the same

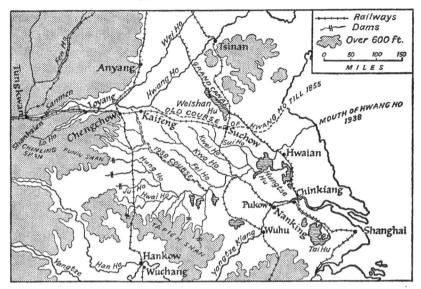

FIG. 52. The Hwai Basin

difficulty is experienced, although in a lesser degree than with the Hwang-ho, of returning the waters to the river bed after flooding.

Because of the devastation caused by the cutting of the Hwang Ho's banks in 1938 and its deflection into the Hwai, the present government has given priority to water conservancy in this basin, even over the Hwang Ho. While not as vast a scheme as that described above, it follows much the same pattern. Eight control dams of appreciable size, i.e. 100 ft. or more in height, have been built in the upper reaches of the Hwai and its tributaries, large retention reservoirs have been constructed in its middle reaches, hundreds of miles of dykes have been raised and strengthened and the out-let from the Hungtze Hu canalized and controlled.

The plain is crossed from north to south by the Peking-Tsinan-Hsuchow-Nanking-Shanghai line and the Peking-Chengchow-Hankow line. The Grand Canal, built by Kublai Khan, fell into disrepair under the later Manchus and is now being dredged and largely rebuilt.

1g. *The Shantung Peninsula*

Although Shantung (Eastern Mountains) province physiographically belongs rather to the Liaotung peninsula and North Korea, it is convenient to deal with it in conjunction with the Hwang Ho as it is so intimately integrated with Chinese history and philosophy. It was here that early cultures flourished; here that the early emperors sacrificed to Shang-ti, Lord of Heaven; here that Chinese philosophy found its home; and here that Confucius was born and later buried. Here also was the home of the Kung dukes and it still is the home of their direct descendants. A further reason for including the Shantung Peninsula in the Hwang Ho basin is that it divides the North China Plain from the Hwai plain and is the bastion against which the Hwang Ho has shifted its course, now north, now south through the centuries.

The province consists of the denuded remains of an ancient mass of Archaean rocks, having a north-east south-west trend. It is the southern extension of the East Manchurian Mountains of Liaotung, and is divided into two halves by a depression which runs from north to south. The eastern portion is made up mainly of Archaean schists and gneiss with some crystalline limestone. The western half, which is higher, is largely of carboniferous limestone and contains the main coal measures. It has been uplifted and very heavily faulted. W. Smith describes western Shantung as a shattered horst.[1] From the coalfields along its western border more than 10 million tons of bituminous coal were produced in 1944. Considerably more than this amount is now being mined. The country as a whole has a landscape of gentle upland. Its highest peak, Tai Shan, rises to a little over 5,000 ft. and is one of China's five sacred mountains. Its slopes are covered with temples and monasteries. The thousands of pilgrims who visit it each year approach it from the city of Tai-an in the west.

The coast of the peninsula is precipitous, rocky and beautiful, but it is dangerous to shipping on account of the frequency of fogs. It has a number of excellent natural harbours which unfortunately have no extensive hinterland. Being so much enwrapped by the sea, its climate is rather more temperate and equable than that of the North China Plain.

Generally the soils on the hills are poor and thin, with the result that severe soil erosion has followed persistent deforestation. This has been met on the lower slopes by extensive stone faced terracing.

The main population, which numbered 48,876,548 in the census of 1953, is second only to that of Szechwan. It is concentrated in the fertile valleys in a density which rivals that of Chengtu. The land is most intensively cultivated but yet is unable to sustain the numbers. Consequently this has been the region from which Manchuria has drawn its seasonal labour and, in more recent years, its main supply of colonists. Crops are

[1] W. Smith, *Coal and Iron in China* (Liverpool 1926).

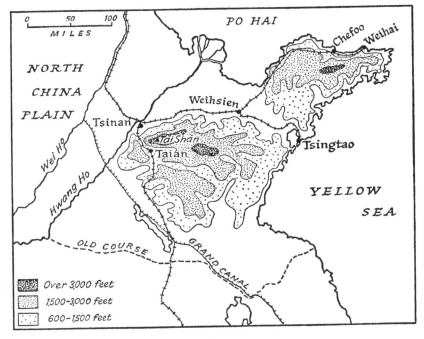

FIG. 53. Shantung

similar to those of the North China Plain. In addition, sericulture has been important. Shantung silk is produced from worms whose food is largely oak leaves rather than mulberry.

Tsinan, the capital of the province, lies on the North China Plain to the west of the Tai Shan and near the banks of the Hwang Ho. Sited as it is in the coalfield, it is rapidly increasing in industrial importance. It is also a railway junction on the Peking-Shanghai railway line. A branch runs west to Wenhsien, thence south-east through the depression to Tsingtao on Kiaochow Bay.

In 1897 the murder of two German missionaries in Kiaochow was made the justification by Germany for the seizure of the region around the Bay. A treaty was signed handing over the land on a ninety-nine year lease. The area was then developed with great energy. A great deal of German capital was sunk in the construction of the port of Tsingtao, in afforestation which changed the landscape, and in building the branch railway mentioned above. Kiaochow Bay, which is 19 miles long by 15 miles broad, is not very deep and required a good deal of dredging. The Japanese took the port in 1914 at the beginning of the First World War after which time trade began to decline. The port was restored to China in 1922.

The main port on the north coast is Chefoo. This was opened to foreign shipping in 1862 after the signing of the 1858 treaty with Great Britain.

2. THE YANGTZE KIANG BASIN

The Chinese words *Ho* and *Kiang* are both used for rivers of major proportions. When used without a prefixing word, *Ho* is taken to mean the Hwang Ho and *Kiang*, the Yangtze Kiang. In fact the Hwang Ho in all early writings is referred to simply as *Ho*. As is often the case with great rivers, the Yangtze has different names for different reaches. The 1,000 miles or so from mouth to the gorges at Ichang are known as the *Ch'ang Kiang* or Long River. From Ichang to Sui-fu it is called *Ch'uan Ho* and above this, in the torrential mountain course, the *Kin-sha-Kiang* or the River of Golden Sand. It is 3,494 miles (5,590 km.) long and has a catchment basin of 714,000 sq. miles (1,850,000 sq. k,m.), more than twice that of the Hwang Ho. It flows entirely south of the Tsinling line and consequently its basin contrasts strongly with that of the Hwang Ho. Both rivers have upper reaches in the Tibetan plateau, but once descended from there the Hwang Ho flows through desert and loess and a natural vegetation mainly consisting of prairie grass, while the Yangtze cuts its way through basin after basin of what was formerly dense temperate and sub-tropical forest.

The river can be divided conveniently and clearly into four main regions: the torrent course of the Tibetan highlands; the Red Basin of Szechwan; the huge middle basin of Hukwang, which embraces the entire area of Hupeh, Hunan and Kiangsi; and finally the deltaic region of Anhwei and Kiangsu.

2a. *The Upper Reaches*

It will be noted that the Kin-sha-kiang, the Mekong (Lantsang in China) and the Salween (Nu Kiang in China) follow closely parallel north-south courses through the length of their passage in the mountainous eastern edge of Si-Kang. At one point only 30 miles separates their beds. They cut deep narrow valleys or gorges of 2,000 to 4,000 ft. deep. As often as not the rivers themselves alone fill the valley bottom. Their steep sides are heavily forested and the valley bottoms are hot, damp and malarious. The rivers are rapid and quite unnavigable.

Four different explanations of this parallelism have been advanced by reputable geologists and geomorphologists. Lee suggests that it is the result of parallel, consequent drainage following sag-lines on a fluted surface of warping; Heims, that it is the Himalayan revolution in Alpine folding; and Credner suggests that channels have developed along parallel belts of weak rock on a peneplain. A final suggestion is that of rifting along fault lines, which form natural spillways for the melt water of Si-Kang glaciers.[1]

These parallel mountain ranges and deep river beds form the borderland between Burma and China, and account for the fact that there has been so little intercourse between the two countries. China has made several attempts at the invasion of Burma in the course of its history, but even if successful, the conquest has been short lived. Communication between the two countries is too difficult to maintain. Kublai Khan sent an expedition of half a million strong to subdue Burma. He lost half his army by disease and exhaustion in crossing these malarious, forested parallel ranges of the border. Wise rulers, if they have concerned themselves at all with Burma, have been content with nominal submission and token tribute.

At approximately 27° N. the Yangtze suddenly reverses its course to NNE and thenceforward zigzags its way in a series of transverse and longitudinal valleys in a general easterly direction for about 500 miles, still flowing in gorges of several thousand feet deep, until it emerges into the Red Basin. In this section it receives a number of large left bank tributaries which conform to the general north-south drainage pattern, but there are no tributaries of note coming in on the right bank. As a result of uplift in the west in the Middle Pliocene, the headwaters of a river (later to be known as the Yangtze) draining into the Red Basin cut back and captured the upper courses of the Red River and part of those of the Mekong[2] (see Fig. 54 overleaf). The effects of this dramatic river capture, bringing as it does a considerable additional flow to the Yangtze from the melting snowfields and glaciers of Si-Kang, are great, both geographically and economically.

Population in this region is very sparse. Hsi-fan hill tribes occupy the wider parts of river valleys in the Si-Kang mountains. Above Suifu (Ipin), in north-west Kweichow, the main concentration of the Lolo, earlier occupiers of Szechwan, is found.

[1] G. B. Barbour, 'The Physiographic History of the Yangtze', *The Geographical Journal*, Vol. 87, 1936.
[2] ibid.

2b. *The Red Basin of Szechwan*

The Yangtze emerges at Suifu from this rugged mountainous region into the hilly, butte country, which characterizes the greater part of Szechwan and which forms the Red Basin, so named by Richthofen. This basin has well defined margins of fold mountains, raised at the end of the Cretaceous. On the north and north-east are the Mitsang and the Ta Pa Shan, behind which lie the Tsin Ling. To the west the steep edge of the Azure Wall Mountains with many peaks rising 17,000 ft. and some to well over 20,000 ft., such as Minya Konka (24,900 ft.) form the rather abrupt border of the Si-Kang plateau. The southern boundary is marked by the mountains of the Kweichow plateau.

From late Cretaceous times until probably the Middle Pliocene this basin formed a lake in which deep deposits of red sandstone were laid down on heavily folded limestone in which the Rhaetic coal measures lie. The red sandstone has since remained largely undisturbed by folding. The uplift in the west in the Middle Pliocene led to considerable rejuvenation and cutting back of the rivers, the capture of the headwaters of the Red River and to a big increase of inflow into the lake. It was probably at this time that the lake found its outlet to the east into the Middle Yangtze depression and so was drained.

The Basin itself slopes generally from about 3,000 ft. in the north to 1,500 ft. in the south. The rivers, which conform to this slope, cut deeply into the soft red sandstone, often exposing the harder limestone and the coal measures. The Chinese characters for Szechwan are 四 川 , meaning Four Rivers, which are usually taken to be the Min, the Lu, the Fu and the Kialing. They all rise in the mountains of the north, and are fast-flowing and unnavigable except for junks and small river craft. Three rivers—the Fu, Kialing and Chu (Pai)—each have broader valleys of 20 to 30 miles width in their middle reaches, but the valley becomes much more constricted after the junction at Hochuan and remains so until their confluence with the Yangtze. The main river hugs the high ground to the south of the basin and flows fast in a deep-cut trench with many short gorges. The river is barely 300 yd. wide at Chungking, where a difference in level of 70 ft. is experienced between summer and winter. Navigation is difficult between Chungking and Kweichow, but comparatively easy as compared with the reach between Kweichow and Ichang, which constitutes the famous Yangtze Gorges.

Between Kweichow and Ichang the river cuts through lofty limestone mountains with precipitous sides, in a series of magnificent gorges which cut right across the rock structures. Earlier physiographic explanation of their causation was that the river had sought out lines of weakness and had cut down along them, but Barbour finds this unacceptable.[1] He suggests two possible explanations, either that the drainage is antecedent or superimposed. He rejects the former in favour of the latter, pointing out that it is probable that the mountains through which the river cuts were folded long before the present drainage system and, moreover, that the Taling-ho, a tributary of the Yangtze in this stretch, has cut down through five anticlines of strata identical with those of the Gorges without being thrown off course. He supports a theory of superposition from a thinly veneered peneplain on the grounds that the folds are sufficiently ancient (probably

FIG. 54. River capture by the Yangtze Kiang

[1] See G. B. Barbour, op. cit., for a full treatment of the subject.

Yenshan) and that 'maximum summit levels show a consistency of skyline across truncated structures'.

The Red Basin is strongly mountain-bound and the Gorges provide the only direct line of entry from the east. The current is very powerful, running at 4 to 6 knots, and in a few constricted and very dangerous stretches it reaches as much as 10 knots. Navigation is therefore notoriously difficult. For a thousand years and more junks have made the passage, hauled up against the stream by tracker teams, sometimes 100 strong, straining in their harness along the specially cut narrow paths high up on the cliff face. The passage is dangerous both for crews and teams and demands a high degree of co-operation, mutual reliance, courage and endurance. Carelessness or faint-heartedness can easily lead to crew or team or both being swept away. In this century specially designed high-powered river steamers have been constructed to contend with this current, but at least until recently, their share of the total river traffic over this reach of the river has been very small.

To meet this difficulty of navigability and also to control floods in the lower basins, there is a great project, still in the survey stage, of building a 400 ft. dam at Sanhsia, near Patung, at the lower end of the Wushan Gorge. If and when this dam is built, it will create a lake reaching back nearly to Chungking. It will generate 15 million kW. and will make the river navigable for 10,000 ton vessels up to Chungking. Two further dams are planned, one at Chungking and the other at Suifu (Ipin).

The climate of Szechwan is startling and unexpected.[1] Here in the heart of a great land-mass, where one might expect to find considerable extremes of temperature, it is equable. Chengtu, which is 1,000 miles inland and 1,560 ft. above sea-level, has an average January temperature of 44°F. as compared with Shanghai, which is 33 ft. above sea-level and on the coast and has an average January temperature of 38°F. Average July temperatures are high, being 78°F. Even so they are 2° less than Shanghai. Very seldom is there snow or frost except on the mountains. The basin enjoys a growing season of eleven months. H. L. Richardson says of it: 'to come to Szechwan in the winter time from the adjoining province of Shensi is to experience a dramatic contrast between the frozen yellowish land of the Wei valley and the lush green growth of the Red Basin'.[2] The reason for the equable nature of the climate lies in the fact that the basin is land-locked by high mountains, particularly on the west and north. Thus the

[1] Chang Pao-Kun, 'Climatic Regions of Szechwan Province', *The Meteorological Magazine*, Vol. 15, Nos. 3–4, 1941.

[2] H. L. Richardson, *Soils and Agriculture of Szechwan* (Chungking 1942).

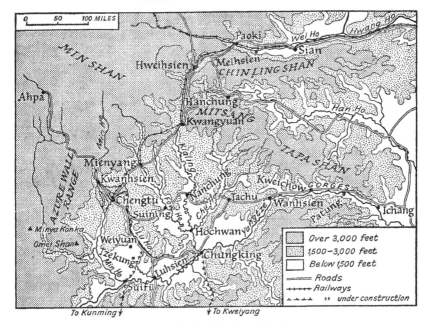

FIG. 55. The Red Basin of Szechwan

bitter winter winds from the Mongolian high are deflected farther east to the middle and lower basins of the Yangtze.

The average annual rainfall for most of the province is 40 in. Winter precipitation is sparse but its meagreness is compensated to some extent by high humidity and cloudiness, which prevents much loss by evaporation. There is a Szechwan proverb which runs: 'When the sun shines the dogs bark'. South of Szechwan lies the province of Yunnan, the meaning of which is 'South of the Clouds'. The heavy summer maximum of rainfall results in a big rise in river levels.

While the natural vegetation for the whole area is forest, the nature of the forest varies considerably with altitude. Much of the valleys and lower hillsides has been cleared and terraced for cultivation. Where left untouched, evergreen broadleafed forest abounds. As is almost universal in China, around each temple and near each village there is a grove of trees, intimately connected with *feng shui*. In Szechwan these groves are composed largely of cypress, golden chestnut and bamboo. Above 2,000 ft. the evergreen forest begins to give way to deciduous broadleafed forest and *Cunninghamia*. At 6,000 ft. little deciduous vegetation remains, coniferous forest, rhododendrons and shrubs and flowers taking its place. Szechwan has achieved botanical fame in that a species of conifer, *Metasequoia*, believed to be extinct, has been found to be growing here. Above 9,000 ft. there are alpine pastures. Purple brown forest soil and mountain pod-zolized yellow soils are general.

Because of the long growing period, more than two thirds of the culti-vated land is double cropped. There is a great amount of terracing and about three quarters of the land is irrigated. The incidence of crop failure is very much less than in the lower basins of the Yangtze. This is due partly to there being less liability to flooding.

It is said that anything grown anywhere in China can and is grown in Szechwan. With the exception of the truly tropical produce of Hainan and south Kwangtung, this is virtually true. Rice is the summer cereal and wheat the winter variety. In addition corn, rapeseed, sesame, kaoliang, soy-beans, sweet potatoes and tea are all grown extensively. Szechwan corn is famous, that produced in the hills being especially good. Opium was formerly a very important cash crop. Many kinds of fruit are grown, including peach, apricot, persimmon, pear and citrus of many varieties. Sugar cane, introduced from Fukien, is grown in large quantities in the river valleys of the south-west and forms an important cash crop. Tea is grown mainly in the north and north-west where winters are warmer and soils good. It is of special quality, the long, round, thick leaf being good for brick tea, which is exported mainly to Tibet. The province stands third to Chekiang and Kwangtung in sericulture. Wanhsien in the east and Loshan in the south are the main producing centres, while the silk trade is centred in Chungking. The *tung yu* (wood oil) tree abounds in the eastern part of the Red Basin, especially in the Gorges and neighbouring areas. There are two main species: *Aleurites fordii* and *Aleurites montana*. The trees flourish on the hillsides up to about 2,000 ft. Szechwan has a wealth of medicinal herbs both wild and cultivated, which have a nation-wide reputation. The main pharmaceutical centres are Kwanhsien, Suifu (Ipin) and Hochuan.[1] Szechwan has a pig population estimated at nearly 16 million and stands first in China as a producer of hog bristles. In 1936 16,000 *piculs* (1 *picul* = 133 lb.) were produced, the trade being centred in Chungking and Wanhsien.

Szechwan's mineral resources lie in coal, salt and oil, and to some extent in iron. Her coal reserves are estimated at 293 million tons of anthracite and 3,540 million tons of bituminous coal, a considerable amount, although meagre as compared with Shansi and Shensi. The main reserves lie in the east and centre but there are outcrops all round the margin of

[1] E. H. Wilson, *A Naturalist in Western China* (London 1913).

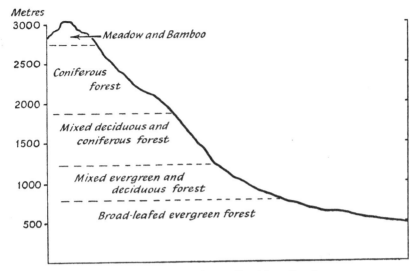

Metres

Meadow and Bamboo

Coniferous forest

Mixed deciduous and coniferous forest

Mixed evergreen and deciduous forest

Broad-leafed evergreen forest

FIG. 56. Natural vegetation at Omeishan, Szechwan

the Basin. Anthracite deposits lie in the west along the lower Min River. Rivers have cut deeply into the red sandstone and have exposed the underlying coal seams, which are often mined by adits. Production in 1939 was 90,000 tons of anthracite and 2,462,159 tons of bituminous coal. This has been stepped up considerably in recent years, but exact figures are not available.

Surveys to date have not revealed any great wealth in iron, and for this reason the present government early scheduled the province as a region of light industry. Iron reserves are estimated at about 2,200 million tons or approximately 1 per cent of the national total. These reserves are located mainly in the south-west of the province, around Weiyuan.

The existence of oil and natural gas in Szechwan has been known for well over 1,000 years. It is not surprising, therefore, that Szechwan has been regarded in this century as one of China's most hopeful fields. Nevertheless it has remained largely unsurveyed until recently. The National Geology Survey of China placed reserves at nearly 400 million barrels, but production before the Second World War was a mere 92,000 *catties* (1 *catty* = 1⅓ lb.) from natural wells. There was a low temperature coal distillation plant in west Szechwan, which produced diesel oil and petrol substitute with a high octane number and therefore good for aviation, but production was very small. Post-war oil survey reports in Szechwan remained disappointing until 1958, when dramatic discoveries were made in

237

the Nanchung region of the Kialing River valley. Reports from the Szechwan Petroleum Survey Bureau state that oil in quantity over a wide field has been found in the Lower Jurassic (1,000–1,500 ft.) and Triassic (6,000 ft.). The oil is of good quality, having high condensability and low viscosity. Many wells have been sunk, producing a number of gushers. If the present promise is sustained, the importance of this new oilfield will be immense.

Salt production, which is an old industry, is centred mainly in Tzeliutsing, which lies between the Min and Lu rivers, although other districts are also engaged in the trade. Brine is raised from deep wells by means of bamboo pipes and is evaporated. Half a million tons of salt are produced annually.

Szechwan has the highest population of all the provinces with 62,303,999 persons, and a density of 1,610 per sq. mile in the crop area.[1] This is in spite of the fact that the population was decimated in a great uprising four centuries ago. The unoccupied land was quickly repeopled by migrants from neighbouring provinces.

Communications in the Red Basin are difficult. The many rivers have cut deeply into the soft red sandstone, rendering cross-country travel difficult. The rivers themselves are fast flowing and navigable only for small craft for the most part. In the past very little wheeled transport has been used, for the 'roads' have been mere flagstone paths. One such road is worthy of mention. In the Later Han (AD 25–220) an official route ran from Chengtu via Mienyang and Kwangyuan, thence over the Tapa Shan and Tsin Ling Shan to Paoki in the Wei valley and so to Ch'ang-an (Sian). This very difficult route was the more remarkable because nearly one third of its length was built on trestles in the precipitous mountain slopes or along stream beds.[2] Along this same route a railway has now been constructed—a really remarkable engineering feat. By means of this line Chungking is now linked to Paoki via Chengtu and to the Lunghwa east-west line along the Wei valley.

The Sino-Japanese War (1937–45) had considerable influence on road communication in the province. The Government (Kuomintang) retreated before the Japanese advance and took refuge behind the mountain walls of Szechwan. During these years a sizeable network of motor roads—much of indifferent surface—was built. This has been further developed since 1949.

[1] Lossing Buck, *Land Utilization in China* (Shanghai 1937).
[2] H. J. Wiens, 'The Shu Tao or Road to Szechwan', *Geographical Review*, Vol. 39, 1949.

There is one region of the Red Basin which deserves special mention, and that is the Chengtu Plain. The Min River emerges from the Azure Wall Mountains on to the plain at Kwanhsien, where its flow is checked. The Min is a considerable river, even in winter, being 50 yd. wide and 6 ft. deep. In summer it is a torrent nearly half a mile wide. In earlier times it fanned out over a wide area, causing a stone waste in the north and marshlands in the south. We have seen (p. 70) how Li Ping in the Ching dynasty (221–206 BC) controlled this flood and converted the area into one of the most fertile lands. The rice crop from the Chengtu plain is reckoned at about 4–5 tons per acre, nearly twice the average elsewhere. It is now the most densely populated agricultural region in the world. Perhaps what is even more remarkable than this transformation from waste to productivity is the fact that for more than 2,000 years the great network of channels and sluice-gates has been maintained through all vicissitudes. Nowhere else in history is there such an example of material and organizational continuity.

2c. The Middle Basin

When the Yangtze emerges from the Gorges at Ichang it descends on to a series of three plains, which, since they lead into one another and are of much the same geographical nature, we have grouped under the heading Middle Basin. These plains are the successors to former depressions or old lake basins, which have since been largely filled with eroded red sandstone from Szechwan. In many parts there are thick 'red beds' lying unconformably on limestones, micaceous sandstones, quartzites and conglomerates. The process of in-filling these depressions is not yet complete. Much of the land is lake-studded along the Yangtze borders and along the Han near its confluence with the mainstream. The flat or gently undulating plain is often suddenly interrupted by isolated hills or ranges of hills rising abruptly. Characteristic are the well known Lushan Hills, near Liukiang, which are 8 miles by 15, springing from the plain to over 5,000 ft.

In its course through the Middle Basin, the Yangtze or Ch'ang Kiang, as it should be called in these lower reaches, receives its biggest tributaries. Its great left bank tributary, the Han, rises in the Tsinling Shan and flows for nearly 300 miles in an easterly direction between the Tsinling and the Ta Pa Shan. On entering Hupeh it turns south in a much broader valley or flood plain and widens its bed, which varies between half a mile and a mile in width over much of this reach. It turns abruptly east again and threads its way through a maze of lakes as it approaches its confluence with the Yangtze at Hankow (Mouth of the Han). Two large right bank tributaries, the Yuan and Siang, and many smaller rivers empty into the

Tung Ting Lake before it, itself, enters the main river. The Han and these right bank rivers constitute the catchment area of the first of the three plains and are known collectively as Hukwang, i.e. Hupeh and Hunan, which names mean North of the Lakes and South of the Lakes, respectively.

About seventy miles below Wuhan spurs of the Ta Pieh Shan on the north and the Wan Fu Shan on the south approach the Yangtze at a point known as Split Hill. Here the river narrows to less than half a mile in width. Split Hill marks the point of entry into the second of the plains of the Middle Basin, the catchment basin of which constitutes practically the whole of the province of Kiangsi. This plain is drained by the River Kan, which empties into the Poyang Lake, thence into the main river.

Near the port of Anyang the Ta Pieh Shan approach the river, which again becomes more constricted. The place of narrowing is known as 'Hen Point' and leads into the third of the Middle Basin plains, which occupies most of central and southern Anhwei. Its boundaries are lower and less well defined than the two to the west. The eastern limit is marked by a further narrowing of the river at 'The Pillars'. From this point to the sea the land assumes a true deltaic character.

The river bed at Ichang is only 130 ft. above sea-level and is 960 miles from the sea. In winter it flows slowly in a channel sometimes barely 6 ft. deep at Hankow. At this time of year a trip from Hankow to Shanghai, even on the large river steamers, is dull, since one sees merely the high brown mud banks, except where the hills come down to the river. In summer the scene is entirely changed: the river comes down as a mighty flood. At Hankow, where the Yangtze is nearly a mile wide, the average difference between summer and winter levels is 45 ft. In times of exceptional flood, as in 1931 and 1954, the flow reaches astronomical figures. In 1954 the flow was measured at more than $2\frac{1}{2}$ million cu. ft. per second (cf. the Thames, 2,300 cu. ft. p.s.).

Throughout the length of the plains of the Middle Basin there is a network of dykes which serve to keep normal summer flood waters from wide tracts of arable land, but over large areas flooding is a usual annual occurrence. The many lakes which cover the land adjoining the river are thereby linked up. Marco Polo and Abbe Huc were amazed at the size of the river. Both reported it as being more than 10 miles wide and both were discredited on that account. But both saw it in the summer when river and lake combine and when it is, in fact, that width in some places.

This variation in régime constitutes a difficulty to navigation at the height of both winter and summer seasons. In winter the shallow water and shifting sandbars make the passage up to Wuhan in craft drawing more than 6 ft. most difficult, necessitating almost constant sounding. In summer, vessels of 10,000 to 15,000 tons can reach Wuhan. Navigation dangers at this season consist mainly in keeping to the river channel and in not wandering off over the countryside as smaller craft are liable to do. If grounded and the river level falls suddenly, as it often does, such craft may be left high and dry for a whole year. For many decades the river course up to Wuhan has been marked by lightships and light buoys in the summer.

While not as subject to flood as the Hwang Ho, the Yangtze nevertheless experiences at times serious inundations in the Middle Basin. The load of silt brought down is heavy: it is estimated that 5,000 million cu. ft. of solid material passes Hankow every year. However, the flow and scour is sufficient to carry the greater part of this down to the sea. Thus the bed of the river has not been built up above the level of the surrounding land. But the Middle Basin is a large region of heavy summer rainfall and the normal inflow from the large tributaries is very great. The two great lakes Tung Ting and Poyang, which in the past have acted as reservoirs and so helped to regulate the flow, are now shallow and less effective for this purpose. Much of the rainfall comes from cyclones which pass down the Yangtze valley. When, as happened in July 1931, a series of seven cyclones passed in quick succession, the flood waters rose to record heights. Then 53·6 ft. were recorded at Hankow, topping the Bund by 6 ft. Nearly 35,000 sq. miles, much of which was under crops, were flooded. Another similar, even greater, rise was recorded in 1954. While big cities like Wuhan were saved by raising the local dykes, disastrous floods again covered the countryside.

The long-term scheme to meet this menace, as we have seen, is to build great dams higher up the river in the Gorges and at Chungking and Suifu and so control the flow. Immediate measures have taken the form of constructing two large artificial retention basins, one at Shasi between the Yangtze and the Tung Ting Lake, and the other at Tachiatai between the Yangtze and the Han, the latter being for relief of the Han's flood waters. Dykes have been raised to form a reservoir or basin into which the waters can be deflected and held when the river is in spate and is threatening to burst its banks. The Shasi basin is 355 sq. miles in area and has 7,000 million cu. yd. capacity. It was built in 1954 in 75 days, using very little machinery and an army of 300,000 workers. It is used for cultivation in the winter months but held in readiness during the summer.

Winters in the Middle Basin are short and cold. About 400 miles separates north Hupeh from south Hunan and consequently a considerable

Flow at Shasi in spate =
2,100,000 cu. ft. per sec.
River capacity = 1,600,000 cu. ft. per sec.

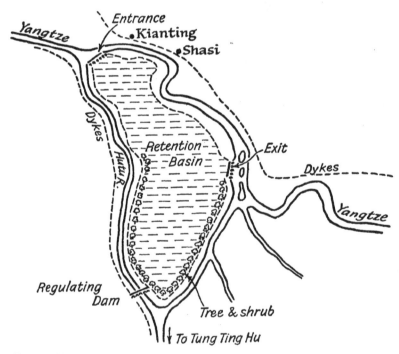

FIG. 57. Yangtze water conservancy: Shasi and Tachiatai retention basins

difference in temperature is experienced. Average conditions are reflected in the climatic statistics of Hankow:

Average	Jan.	Feb.	Mar.	Apl.	May	Jun.	Jly.	Aug.	Sep.	Oct.	Nov.	Dec.	Total
Temperature (°F.)	40	43	50	62	71	80	85	85	77	67	55	45	
Rainfall (inches)	1·8	1·9	3·8	6·0	6·5	9·6	7·1	3·8	2·8	3·2	1·9	1·1	49·6

Winters are wetter than on the North China Plain. Occasional medium falls of snow occur and rarer, beautiful but rather destructive glazed frosts. Summers are hot. The summer rains and the vast flooded paddy fields produce a very high relative humidity, which is most enervating. The hot damp nights with temperatures over 80°F. and a relative humidity of over 85 per cent unrelieved by the slightest breeze have to be experienced to be appreciated. Rainfall has a better seasonal distribution than north of the Tsinling Shan. The growing season varies from nine months in the north to ten months in the south.

Soils in northern districts tend to be neutral. Acidity increases quite appreciably in the south.

The dense mixed forests of broad-leafed trees and *Cunninghamia*, which covered these plains 2,000 years ago, have long since been cleared as have also the hills lying in their vicinity. The wilder and more mountainous borders in west Hupeh and south-west and south Hunan are still heavily forested, although cutting has been very heavy during this century. It is from these regions that the huge timber rafts of pine, fir, bamboo, locust, maple and camphor are floated down to Wuhan and beyond to the cities of the lower Yangtze.

The Middle Basin is one of the most productive regions of the country. Wherever possible rice is grown as the summer cereal and it is certainly the most important crop. The region around the Tung Ting Lake and in the lower Yuan and Siang is renowned and is in fact the most important rice area in China. The Kan plain and the lower Han are also heavy producers. So keen is the desire to grow rice that farmers in the upper Han persist in its cultivation when wheat would be the more reliable crop and sometimes they meet with disaster. Cotton is also a very important summer crop, followed by corn and soy-beans.

The chief winter crops are barley, wheat, rapeseed and sesamum seed and broad beans. Barley is preferred to winter wheat as it matures and ripens early enough for the rice crop to be planted. About two thirds of the cultivated area is double cropped and irrigated. Much of the irrigation is by gravity from ponds at the heads of valleys. Where water has to be raised, reliance is placed on wooden paddle pumps operated by human power. Although mechanization in irrigation is the objective, it will be some time before it is the rule rather than the exception.

Tung yu (wood oil) is an important product. The trees are grown mainly in west Hupeh and in the upper Han and Yuan river valleys. The oil is expressed locally by crushing and it is sent down to Hankow in bamboo crates lined with oiled paper.

Tea is grown mainly on the sides of the rolling hills south of the Tung Ting and Poyang lakes in Hunan and Kiangsi. While its place in international trade has waned since the 1880's, when Indian and Ceylon teas became popular in Europe, it is still a very important crop today. Tea remains the universal drink of the Chinese people.

Cotton is an important Hupeh crop in the middle and lower Han basin. It is mainly short staple of less than 1 in. Wuhan is a very active textile manufacturing centre. No. 1 Cotton Mill at Wuchang was established at the turn of the century. New mills equipped with the latest modern machinery have recently been established in Hanyang.

Of the fibres grown in the Middle Basin, ramie, from which grass cloth or Chinese linen is made, is outstanding. It is a perennial plant which grows on the hillsides. A fair amount of hemp is harvested and some jute is grown south of the river at Wuchang.

The mineral wealth of this region consists mainly of the high grade iron ores of the Tayeh district of Hupeh, the coking coals of the P'inghsiang district of Hunan and the rich, newly discovered Onan field at Puchi. The development of these resources at the turn of the century has been described in the section dealing with economic development before 1949. Since 1949 new deposits near to Tayeh have been found and four new mines opened at Cheng Chiao, Lin Hsiang, Jin Shan Tien and Lung Liu Shan. Modern mining equipment and an ore dressing plant, using froth flotation and magnetic separation methods to separate iron and copper have been installed. The railroad from Tayeh to Shih Hui Yao, river port for shipping the ore to Hankow, has been electrified and there is now rail connection between Tayeh and Wuchang. Shih Hui Yao itself has local coal and some of the purest limestone in the land. It has developed some smelting, but the bulk of the raw material goes upstream to the great works at Wuhan. Based on the fine supply of limestone and easily available coal, the biggest cement works in this part of Asia were opened at Shih Hui Yao in 1949.[1]

In company with Paotow, Wuhan was selected by the economic planners of 1952 as one of the two new centres for the development of heavy iron and steel industry. The old site of the iron and steel works at Hanyang,

[1] T. R. Tregear, 'Shih Hui Yao—a Chinese River Port with a Future', *Geography*, April 1954.

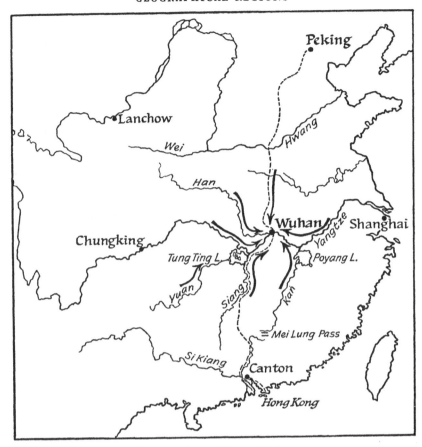

FIG. 58. The nodality of Wuhan

chosen by Viceroy Chang Chih-tung, has been abandoned to modern textile works and vast new development has taken place at Wukang (Five Mile Creek), five miles below Hankow on the left bank of the Yangtze. Here the Wuhan Iron and Steel Corporation has built a fully integrated plant, including ore-dressing and ore-sintering plant, blast furnaces, open hearth ovens, rolling mills, refractory materials plant, etc. In all there are 18 main and 31 subsidiary workshops, covering 3·86 sq. miles (10 sq. km.). The blast furnaces are the largest in this part of Asia and are entirely automatically controlled. They have an annual output of 3 million tons of steel. One blast furnace has an output of 2,000 tons of pig-iron per 24 hours. The blooming mill can roll 7 and 15 ton steel ingots and is moreover entirely automatic.

FIG. 59. Wuhan's iron and coking coal supplies

Wuhan, which is the collective name given to the three cities of Hankow, Hanyang and Wuchang at the confluence of the Yangtze and the Han rivers, lies in the heart of the Middle Basin. It provides a classic example of nodality. The Yangtze itself forms a great east-west highway from the Red Basin of Szechwan to the fertile plain of Kiangsu. The Han River communicates with the north-west over the Tsinling Shan to the Wei valley and Lanchow. The Yuan and Siang link Wuhan with Kweichow in the south-west and Kwangtung in the south. Furthermore, Kwangtung is linked with Wuhan via the Mei Lung Pass, thence by the Kan River and Poyang lake. This, incidentally, was part of the imperial route between Canton and the capital.

In the course of its 1,700 years of history, Wuhan has been sacked many times but has always risen, phoenix-like, from its ashes thanks not only to its nodality but to the fact that the whole region is so productive. The twentieth century has accentuated its centrality by the building of the Pei-Han Wuchang-Canton railway. Until 1957 this railway suffered the severe handicap that all north-south through traffic had to be ferried across the river at Hankow—a difficult task in winter because water was so low

246

PLATE XV: Shelter belts are able to reduce wind speeds from 38 m.p.h. to 15 m.p.h. and are successful in controlling dust storms and sand movement. Wherever possible fast-growing varieties (fir, larch, poplar and camphor) are planted, as they are valuable later as timber. This network is in Honan.

PLATE XVI: One of the most important factors contributing to the vast material development is the rapidity with which the Chinese have acquired the skills and techniques of modern industry. This shows part of the Peking State No. 1 Cotton Mill.

PLATE XVII: Between 1955 and 1960 the annual road construction leaped from 12,600 miles to 40,000. This section of new road is laid on the site of the old city walls of Hofei.

PLATE XVIII: The Yangtze River bridge (1,810 yards long) carries a lower double railway track and an upper six-stream vehicle and pedestrian highway. It is an engineering feat of which the Chinese are justly proud.

PLATE XIX: Much of the old Tibet highway has now been built to allow motor traffic to penetrate hitherto inaccessible regions.

and no easier in summer because of the strength of the current. Ever since 1913 there were projects for building a bridge but these never progressed beyond the survey stage until 1955, when construction actually began. With the help of Russian engineers the river has been bridged between Tortoise Hill in Hanyang and Serpent Hill in Wuchang at its narrowest point; albeit the bridge is 1,810 yd. long, of which 1,252 yd. are actually over the river. The bridge carries a lower double rail track and an upper six-stream highway and is high enough to allow the passage of 10,000 ton ships during summer high water. It is an engineering feat of which the Chinese are justly proud. Since then the Yangtze has been bridged at Chungking and plans are advanced for a further bridge at Nanking.

Wuchang, the provincial capital, has been the administrative and educational centre. It was a walled city until 1928 when, after the Kuoming-tang victory, the walls were pulled down and a ring road constructed. Expansion to the east of the city for administrative, educational and industrial purposes since 1952 has been very great indeed. Hanyang, also walled, was favoured as a place of retirement for officials and gentry. Outside its walls at the confluence of the Han and Yangtze, the Hanyang Iron and Steel Works were built.

Hankow has always been the commercial centre but it entered into

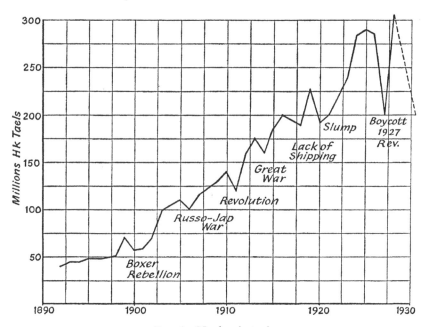

FIG. 60. Hankow's trade

247

international trade only in 1861 when a British concession was granted. In the closing years of the century further concessions were given to France, Germany, Russia and Japan. The nature and the rise of Hankow's trade are shown by the figures below, which serve also to shed some light on production and trade in the Yangtze Basin, particularly in Hupeh and Hunan. The figures are in Haikwan *taels*, a customs unit of 583 grains of silver, a stable measure which is no longer used. Hankow's part in foreign trade rose rapidly, if unsteadily, from 1861 to 1928. Especially in the first quarter of this century political events were clearly reflected in the volume of trade as the accompanying graph shows.

The status of all international ports on the Yangtze suffered a series of heavy blows after the Kuomintang revolution of 1927 from which they have not yet recovered. The loss of foreign concessions, the closing of the Yangtze to foreign bottoms, the world slump of 1932, the Sino-Japanese War, during which Hankow suffered particularly heavy bombing by first Japanese and then American aircraft, and finally the stagnation of trade between 1946 and 1949—all served to bring international trade on the Yangtze practically to a standstill.

Hankow Trade Figures[1]

Foreign Exports 1872		Foreign Exports 1928	
Haikwan Taels		Haikwan Taels	
Tea	12,356,541	Tea	17,002,282
Wood oil	1,384,149	Wood oil	9,615,538
Silk	1,281,884	Sesamum seed	6,053,570
Vegetable tallow	606,763	Cotton seed	1,552,747
		Vegetable tallow	1,815,319
		Vegetable oils	19,037,174
		Egg products	12,174,570
		Tobacco	5,469,875
		Fibres	3,455,656
		Cotton	2,834,777
		Beans	2,800,000
		Flour	2,328,851
		Timber	2,151,396
		Bristles	1,586,665
		Silk (cocoons)	1,007,290

Shipping on the Yangtze may be divided into three categories. Ocean-

[1] From Chinese Maritime Customs Reports.

going ships of 10,000 to 15,000 tons can navigate to Hankow in the summer months and smaller ocean craft can reach Ichang. Specially constructed river boats of 5,000 tons, drawing less than 6 ft., together with large numbers of smaller launches, are used the year round for both passengers and goods, and for long and short journeys. Lastly there are innumerable junks, large and small, which ply up and down the rivers and along the south-east coast, each flying its distinctive flag indicating the nature of its cargo, and each providing a home for one or more families.

Owing to the big change in river level between summer and winter, wharfage at the big river ports is impossible. Landing is made at hulks connected to the shore by pontoon bridges, which can be adjusted in length according to the needs of the season. The smaller towns along the river have to rely on carrying their passengers and freight out in small junks to the passing river steamer in mid-stream.

The main ports along the Yangtze in the Middle Basin are Wuhu, Anking, Kiukiang, Hankow and Ichang, all of which were former treaty ports.

2d. *The Yangtze Delta*

At a point, known as 'The Pillars', about half-way between Wuhu and Nanking, the Yangtze emerges on to a flat, deltaic plain. Apart from the hills which persist for some way along the left bank, the delta is unrelieved by uplands of any height and is reminiscent of the Fenlands of England and the polders of Holland. Much of it is the same level as the river and some parts are below. It is a mass of intersecting canals, many of which serve the treble purpose of drainage, irrigation and communication. From Nanking to the sea there is no clear physical divide between the Hwai and the Yangtze. The former has, on occasion, emptied into the East China Sea through the northernmost of the Yangtze's outlets when that river had three mouths. The southern limits of the delta are marked by rolling hill-land which rises on the Chekiang northern borders into the Pai Chi Shan. The delta extends as far south as the borders of the Hangchow Bay and is covered by innumerable lakes, the greatest and most famous of which is the Tai Hu.

The Yangtze carries a very heavy load of silt. It is estimated that more than 5 billion cu. ft. of solid material passes Hankow each year, but owing to its greater volume and greater velocity than the Hwang Ho, a much larger proportion of this load reaches the sea. Some subsidence is taking place in the delta area, but deposition is occurring at a greater rate and the

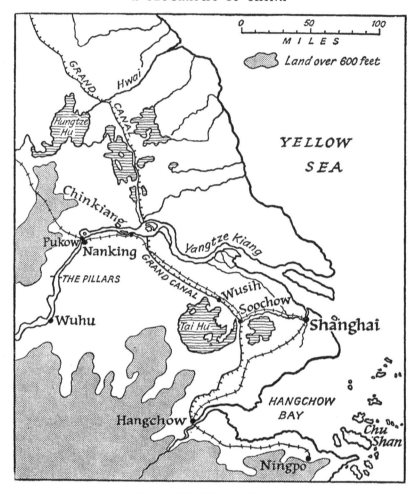

FIG. 61. The Yangtze Delta

coastline is being pushed seaward quite rapidly. Much of the silt is carried southward by the East China Cold Current, which flows south along the coast with the result that the Chusan Archipelago is filling up and Hang-chow Bay becoming shallower. As the silting in the delta takes place, reclamation dykes are built in which there are sluice gates, which allow an inflow during the flow of the tide, bringing further silt, which settles, and also fish, which are netted at the gates as the tide ebbs. The Yangtze is tidal as far as 'The Pillars'.

Contrary to expectation, approach to the seaboard brings no increase in equability of climate. Winters in the delta, although short, are severe. The average January temperature of Shanghai is 38°F. as compared with Hankow's 40°F. The region is open to the cold north winds which sweep down from the North China Plain and the Hwai Basin. Summers are hot, damp and very enervating. Rainfall is better distributed than in the north with an average of 1·3 in. in December and 7·4 in in June. Rainfall variability is much less here than in the northern wheat area. Lossing Buck records 11 calamities between 1904 and 1929 as compared with 24 in the north.

The combination of rich alluvium, a hot, wet summer and a very shallow water table, giving adequate water supply, leads to heavy cropping. Rice is by far the most important summer cereal and stands second in production only to Hunan in the Middle Basin. Other summer crops are cotton, soy-beans and corn. Rice is grown in rotation with winter wheat to some extent, but barley is generally preferred because of its earlier ripening. Rapeseed and broad beans are also winter crops. More than two thirds of the delta is double cropped.

Before the Sino-Japanese war of 1937, silk was the most important cash

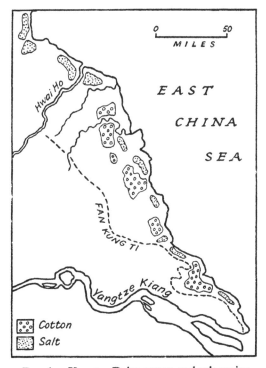

FIG. 62. Yangtze Delta cotton and salt region

crop produced in the delta region. During the early part of the century it had fallen on bad times owing largely to neglect and ignorance of sound methods of sericulture. This was further accentuated by the slump in world trade between 1929 and 1933. To meet this situation schools were opened to improve silkworm rearing in all its stages. Co-operative societies were formed and for a while the industry flourished. Subsequent war and the development of synthetic products, such as rayon and nylon, have dealt a severe blow. Mulberry trees line the dykes and holdings, and mulberry groves dot the whole countryside.

Cotton growing, which is now more important than silk, is concentrated largely in the coastland strip between the mouth of the Hwai and the Yangtze. In the seventh century a long dyke or sea wall, known as the Fan Kung Ti, was built against the inroads of the sea. Since then there has been a steady seaward extension of the land through deposition from the Hwai and the Hwang Ho. A further dyke has more recently been built along the coast. The land nearest the seaboard is devoted to salt production by evaporation in shallow pans, the ground water being highly saline. Farther inland and as the land loses its salinity, cotton replaces salt production. Cotton is more tolerant of salt than most other crops and good yields can be grown on ground where the salinity is not more than 0·16 per cent. The region lent itself to plantation development in the hands of large companies, a few of which had as much as 100,000 acres. These, since the revolution of 1949, have been turned over to state farms or the communes. This low-lying coast is rather vulnerable to flooding, especially during the high winds of the late typhoons.

Population is very dense in the delta. Kiangsu, with an area of 107,300 sq. miles, the coastal strip of which as we have seen is saline and not densely peopled, has a population of 41,252,192. The crop area has an average of 1,360 people per sq. mile.

The delta region is better served by railways than the rest of the Yangtze Basin. The Peking-Tientsin-Tsinan line divides at Pengpu, one branch going to the Yangtze opposite Wuhu and the other to Pukow, where there is a rail ferry to Nanking. It is at this point that another bridge across the Yangtze is contemplated. From Nanking the railway runs to Soochow and Shanghai, thence to Hangchow and Ningpo.

The Grand Canal, keeping to the eastern edge of the Hungtze Hu and Kaoyu Hu, cuts right across the delta to Chinkiang. It is continued through Nanking to Hangchow via Wusih and Soochow. During the first half of this century it fell into disrepair, so that through traffic was not possible. Recently it has been resuscitated and is again in operation. Local communication and transport are very largely by water, using the vast network of irrigation and drainage canals. Retailers from the towns in their floating shops market their goods in the villages and hamlets, and agents, merchants and entrepreneurs collect raw silk and cotton by boat.

The lower Yangtze is rich in towns of note. At the head of Hangchow Bay is Hangchow itself, a city of bridges and canals, held to be one of the most beautiful in the world. Marco Polo visited it about AD 1290 and was ecstatic over its charms. It was the capital of China under the Sungs between AD 1127 and 1278. Soochow, east of the Tai Hu, vies with it in beauty. The Chinese have a saying 'Heaven is above but Soochow and Hangchow are below'. Hangchow Bay is so shallow as to permit no port along its shores. It is subject to very rapid tides; at spring tide a bore of 8 ft. or more sweeps up to its head. Across its mouth lie the islands of the Chusan Archipelago, mountainous and about one hundred in number. The largest, which measures about 20 by 10 miles and gives its name to the group, is a great Buddhist centre.

Nanking, the 'Southern Capital', has a long history as the capital of various kingships from AD 229 onward, but it was not until 1368 that it became, under the Mings, an imperial capital and then only for 34 years. It was famous for its magnificent buildings until the delta region suffered so severely during the Taiping Rebellion. Nanking was taken by the rebels and its monuments were entirely destroyed. Between 1927 and 1949 it was adopted by the Kuomintang as the national capital. Sun Yat Sen was buried there in 1925. Its population in 1957 was recorded as 1,419,000.

Shanghai overshadows all other cities of the delta both in size and in economic importance. It stands at the mouth of the Yangtze and has for its hinterland the whole of the vast Yangtze Basin from which flows the produce of one tenth of the world's population. Its siting, however, is by no means as impressive as its situation. It stands on the left bank of the Hwangpoo or Woosung creek, about 12 miles above its confluence with the Yangtze. The Hwangpoo is the main channel draining the system of lakes to the west. As it approaches the Yangtze it widens to about 700 yd. at its mouth. It is strongly tidal and has a good tidal scour. Below the confluence the Yangtze itself requires constant dredging and most careful conservancy in order to maintain a satisfactory navigation channel. More than 2 million cu. yd. of mud must be dredged annually.

Until Ming times Shanghai was a mere fishing village. Later it developed into a small walled town fortified against Japanese pirates. In 1843, after the Treaty of Nanking, Sir George Balfour chose a site alongside this walled town for a British concession, which in 1863 became the International

Settlement. In spite of the ravages of the Taiping, who sacked the native city, trade grew rapidly and by 1895 its population was 411,573, of which 286,753 was in the settlements and 125,000 in the walled city. It was over 1 million by 1910 and 6 million in 1954. Before the Second World War Shanghai stood as the eighth largest port in the world. For a short while after the 1949 Revolution the planners tried to disperse some of this population and to divert some of its industry elsewhere, but apparently this attempt has been given up and its trade and manufactures have resumed their normal trend.

3. THE SI KIANG BASIN

The Si Kiang Basin is divided from the Yangtze Basin by a broad mountain mass, descending in height from the Kweichow plateau and extending eastward in Kwangsi in a general east-west trend in the ranges of the Nan Ling. These ranges veer to a NE-SW direction in Kwangtung and continue this trend right through Fukien to Hangchow Bay. Pressure from the north-west in Palaeozoic 'Sinic' times, accompanied by extensive folding, determined this NE-SW axis.

The inland ranges on the borders of Kwangtung, Kwangsi, Hunan and Kiangsi are mainly old pre-Carboniferous, sedimentary formations, comprising slates, sandstones and schists, with intercalated beds of shales and clays which have been intensely folded and metamorphosed. The ranges generally have sharp crests and are steep-sided, making rice production difficult. It is for this reason that this inland region is comparatively sparsely populated. To the west there are extensive limestone formations, giving rise to a karst topography, which is every bit as fantastic in its grotesque shapes as Chinese art depicts it. Spires and pinnacles crowned with temples rise abruptly from the plain. Caves, caverns, sink-holes, underground streams, intermittent streams—all the phenomena of karst country are to be found here in profusion.

Nearer the coast and along the littoral a great granitic batholith extends from Hangchow Bay to Vietnam. Most of the older overlying sedimentaries, which are the only fossil bearing strata in this area yet discovered, have been eroded. Large outpourings of igneous rock are characteristic of the coastal areas, which have been folded and faulted. Deep submerging, probably in Pliocene and Pleistocene times, has resulted in a rugged ria coast, rich in good harbours along its whole extent. High humidity and high temperatures together produce rapid and deep disintegration of the granite,

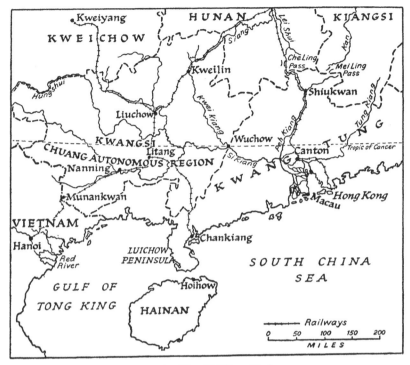

Fig. 63. The Si Kiang Basin

which is then very prone to gully erosion, producing a poor, infertile soil with lateritic qualities. Farther inland derived soils are sandy with a good admixture of clay and are therefore more fertile.

Kwangtung and Kwangsi are drained by three main rivers which unite in one deltaic mouth, the Pearl or Chu River. These three are the Si (West) River itself, the Pei (North) and the Tung (East), of which the former is the most important.

The Si Kiang is much smaller than the Yangtze, being only 1,650 miles long as compared with the Yangtze's 3,494 miles. Nevertheless it is a considerable river both in width and volume. It is half a mile wide at Wuchow where it is joined by the left bank tributary, the Kwei Kiang, and in times of great flood it has a volume of over 2 million cu. ft. per second. Because it flows for the most part through narrow and confined valleys, variations in river level as between summer and winter are also very great. At Wuchow they measure between 70 and 80 ft. There is no vast flood plain as there is in both the Hwang Ho and Yangtze. Only when the river issues out on to the alluvium of the delta just above Shamshui is there any considerable expanse of flat land. Even this is broken especially near the coast, by islands of hills which rise abruptly from the silt.

The Si Kiang Basin embraces most of the two former provinces of Kwangtung and Kwangsi, which were known collectively as Liangkwang —the Two Kwangs—and were governed by one viceroy until the downfall of the Manchus in 1911. Kwangsi is now known as Kwangsi Chuang Autonomous Region. The intention of these Autonomous Regions, of which there are three today in China, is to give some special assistance to and protection of, the rights of the aboriginal or minority peoples in the area. Here in Kwangsi, which has a total population of 19½ million, there are 6·6 million Chuang, 2·6 million Miao and about 9·6 million Yao. The total population of Liangkwang is just over 54 million. Thus the vast majority are Han Jen, or more correctly T'ang Jen, speaking the difficult Cantonese dialect with its ten tones. The Cantonese claim that theirs is nearer the original pure Chinese than Mandarin, the official language and the speech of the north. Mandarin, it is alleged, has been modified and corrupted by repeated invasions of barbarians from the north. It should be noted that the People's Government regards China as a unitary government. The Autonomous Regions are not federated states as in the USSR but are integral parts of a united whole.

The delta, which measures about 70 miles from north to south and 50 from east to west, consists of a network of distributaries on the west and north and a wide main channel, the Pearl River of the east, which empties into the South China Sea between the islands of Portuguese Macao on the west and the islands of British Hong Kong and the New Territories on the east.

The channels and canals of this maze, measuring about 1,500 miles, are constantly changing course: old channels are abandoned and new ones developed. Through the centuries man has worked continuously to maintain the channels from silting, until now it is estimated that there are two to three times as many man-made canals as there are natural distributaries. The farmlands reclaimed here are protected against flooding by dykes lined with trees, in much the same manner as those of the Yangtze Delta.

The Tropic of Cancer passes just north of Canton, cutting the Si Kiang Basin into two more or less equal parts. The northern half, because of its general higher elevation and its more northerly position, is sub-tropical, while the southern half is essentially tropical, especially along the south-west coast and on the island of Hainan. The region experiences three quite clearly marked seasons. From mid-October to mid-January is a delightful time with clear skies, warm and dry. From mid-January to mid-March,

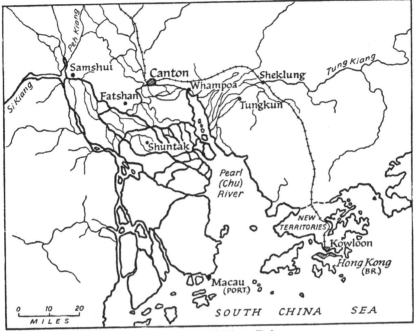

FIG. 64. The Si Kiang Delta

around the time of Chinese New Year, is the coldest period when damp, dull weather is experienced. This is followed by a long, wet, enervating summer when both temperature and relative humidity are high. Walls stream with condensation and shoes grow mould overnight. Only on the higher ground is frost experienced. Annual rainfall ranges between 60 and 80 in., three quarters of which falls in the long hot summer. Temperatures throughout the year are high enough in the entire area to give a twelve month growing period.

	Jan.	Feb.	Mar.	Apl.	May	Jun.	Jly.	Aug.	Sep.	Oct.	Nov.	Dec.	Tota
Kweilin													
Temperature (°F.)	48·4	44·2	49·3	54·1	66·7	73·4	80·1	83·3	81·7	78·8	72·1	59·9	
Rainfall (inches)	1·6	3·9	4·3	9·4	14·1	16·4	8·0	7·0	3·0	2·6	2·1	1·6	74·1
Canton													
Temperature (°F.)	56	57	63	71	80	81	83	83	80	75	67	60	
Rainfall(inches)	0·9	1·9	4·2	6·8	10·6	10·6	8·1	8·5	6·5	3·4	1·2	0·9	63·6
Swatow													
Temperature (°F.)	59	57	62	70	77	82	84	83	82	76	68	62	
Rainfall (inches)	1·4	2·5	3·1	5·6	9·0	10·5	7·8	8·4	5·5	2·9	1·6	1·5	59·7

Such a climate should ensure a vast covering of luxurious tropical and sub-tropical forest, especially in the more mountainous regions. In fact the whole area has suffered severe deforestation. Even in the remoter, steeper parts of Kwangtung, which are sparsely populated (an average of 75 persons per sq. mile), the natural vegetation has been destroyed and replaced by scrub and grasslands. Fenzel[1] attributes this spoliation of the natural cover to three main causes. The Chinese farmer, as he moved southward in the early centuries, encountered forest as an enemy, as an obstacle to be cleared, and he has conserved this attitude ever since. Chinese literati have lacked that appreciation of forests which was present in the hard-hunting feudal lords of medieval Europe. Last, and probably most important, the ownership of the land was by small plots, but forests flourish best under large ownership. It may be that, with common ownership of the land, the present ambitious afforestation projects for this area will come to fruition. Characteristic trees are camphor, acacia, silk cotton tree, *Clausena*, palms including the betel palm, and *Pinus massoniana* (China fir).

Temperature and rainfall combine to give Liangkwang a twelve months growing period. Rice is by far the most important crop. Nearly all the rice grown is fresh-water paddy and is double cropped. A first harvest is taken in late June or early July and a second between mid-October and mid-November. The early crop is a different variety from the late. A little brackish water paddy is grown in the newly reclaimed lands along the seaboard and some upland (dry) rice is grown on the hills inland. Both brackish paddy and upland rice take longer to mature than fresh-water paddy and only one crop a year can be secured. A good fresh-water paddy farmer can usually take a catch crop of vegetables or roots between the November harvest and the spring planting. Other important crops are sweet potato, corn, sugar cane, groundnuts, tea and many kinds of vegetables. Fruit grows in profusion in this tropical climate. Prominent are the pineapple, citrus fruits (tangerine, mandarine and pomelo), lychee, longan, papaya, guava, banana and persimmon. Kwangtung's silk production is still considerable and ranks third to Szechwan and Kiangsu. Some indigo is also produced. Soils are acid and need constant liming, and rapid leaching of the soil due to heavy summer rains necessitates heavy fertilization.

Farms in the south are fairly well stocked with water buffaloes for draught work in the paddy fields. Large numbers of hogs are raised, many of which are exported daily down the Pearl River from Canton to Hong Kong. Poultry and duck farming is also important, particularly in the delta area.

Fishing, both inshore and deep sea, is a considerable industry all along

[1] G. Fenzel. 'On the Natural Conditions affecting the introduction of forestry as a branch of rural economy in the province of Kwangtung', *Lingnan Science Journal*, No. 7, June 1929.

the indented coast and in the shallow waters of the Pearl River estuary. Chinese junk-trawlers, drifters and long liners fish the South China Sea. Inshore waters have been badly over-fished in most parts, but farther south they still provide a rich harvest in spite of inroads by Japanese fishing fleets, which are more highly mechanized and better equipped than the Chinese. The main catches are yellow croaker, golden thread, white herring, mackerel and pomfret. Fresh-water fish farming in artificial ponds, particularly in the delta region is an important and growing side of the fishing industry. The upper Pearl River is particularly important as a breeding ground for fish fry, which is exported to all parts of South-east Asia.

The Si Kiang Basin is scheduled for development as a light industry area. The reason for this is that there is a deficiency of coal in this region. It is possible that this decision will be changed at a later date if the very high hydro-electric potential of the many fast-flowing rivers is properly developed, for there are good reserves of iron ore in both the valley of the Han (Mei) River and in northern Hainan. Manganese is found in quantity in the south-west and in the Liuchow peninsula, and there are rich deposits of tungsten in northern Kwangtung on the Kiangsi border. At present practically all the light industry of Kwangtung is concentrated in the delta area. In addition to the many very small local refineries, large centres dealing with the rapidly increasing sugar cane production have developed at Shunteh and Tungkun. The main silk filature and weaving centres are at Canton, Shunteh and Fatshan.

The island of Hainan, which is only slightly smaller than Taiwan, is capable of much development. The southern part of the island is very mountainous, the Wuchih (Five Fingers) Mountains, with their five radiating ranges, reaching a height of over 6,000 ft. There are wide plains in the north bordering the narrow Chungchow Straits, which separate Hainan from the peninsula of Liuchow.

Much of the island, especially in the south, is occupied by tribes of Miao and Lu, who are still practising *milpa* or *ladang* migratory agriculture. Standards of life are low and the people are prone to the debilitating diseases of the tropics, notably malaria and hookworm. T'ang Jen and Hakka from the mainland are the main occupants of the northern plains. Here, because the climate is more truly tropical than the rest of China, coconut and rubber are being developed in addition to rice cultivation. There are big deposits of high grade iron ore in the north-west awaiting development.

For communication, reliance is still placed on water transport. The

259

rivers, although fast-flowing, are navigable for junk and launch and these carry a great deal of the inland traffic. The Kwei Kiang, a left bank tributary of the Si Kiang, is connected by a short canal with the headwaters of the Siang River and thus provides a continuous, albeit small, waterway to the Yangtze Basin. The new railway to Vietnam follows these same headwaters to Kweilin, thence to Liuchow, Nanning (capital of the Autonomous Region) and Munankwan at the border. One branch line runs from Litang to the coast at Chankiang and another runs from Liuchow up on to the Kweichow plateau to Kweiyang. Both the road and the Canton-Hankow railway follow the valley of the Pei Kiang and cross the Nan Ling by the Che Ling Pass down into the Lei Shui valley and so to Hengyang and Changsha. The road forks at Shuichow in the Pei Kiang valley, one arm running over the Mei Ling Pass to the headwaters of the Kan Kiang and so into Kiangsi. This is the route of the former famous imperial road to Peking.

Canton is the only really large city of South China: it has now a population of 1,300,000. As we have seen, it figured prominently in the events which led to the First Opium War, 1841-2, when it was the gate of entry for foreign goods. The river was then adequate for ocean-going vessels but now is too shallow to permit large ships to reach it. Whampoa, lower down the river, can take vessels of 10,000 tons and is now Canton's outport.

Canton has been the home of revolutionary movements in this century. The abortive revolt, led by Sun Yat Sen, which was the prelude to the Revolution of 1911, occurred here and is commemorated by a monument which is remarkable for the many stones in its make-up contributed by Chinese in the USA. It was from Canton that Chiang Kai-shek led his victorious army northward in 1926 and it was here that the Communist Canton Commune was set up for a few days in 1927.

Haikow on the north coast of Hainan is a small but active port.

4. THE SOUTH-EAST COAST

The south-east coastal region is that area which lies between Ningpo and Swatow, a land of mountains and steep-sided valleys cut deep into the granites, granodiorites, porphyries and igneous rocks, which are the main geological constituents. The western border of the region is marked by the crests of the Wuyi Shan, which are also known as the Bohea Mountains and from which the Bohea congou tea derives its name. The south-east border is the deeply indented, precipitous ria coast, strewn with islands. It is similar to the Kwangtung coast in formation but is more rugged and it differs from the Kwangtung coast in that the excellent harbours have little or no hinterland comparable with the Si Kiang. A series of short, fast-flowing rivers cut their way down to the sea from the mountainous interior. None of them is of use for communication, except for small junks and sampans, which have to be hauled upstream by trackers.

Of the hundreds of streams emptying into the Taiwan Straits four are worthy of note, each of which has a town and port of some size at its mouth. In the north, in southern Chekiang, Wenchow stands at the mouth of the Wu Kiang. Farther south is the largest of the Fukien rivers, the Min, which has one left bank and one right bank tributary which are comparatively large. The Min cuts a deep valley and enters the sea at Foochow without forming a flood plain. About 140 miles to the south is Amoy at the mouth of the Kiulung Kiang, which, while much smaller than the Min, has gentler gradients and a plain nearly 10 miles square above its entry into the sea. Amoy stands on an island and has an excellent harbour. Ten miles to the east and still in the mouth of the bay, Hsiamen Wan, is the island of Quemoy—*Chin Men* in Mandarin and meaning 'Golden Gate'—fortified and held at present by the Kuomintang forces of Taiwan. Thirty miles to the north-east is the modern village of Chuanchow, which stands on the site of old Zayton, the great medieval Chinese port for India, Java and Japan, and much visited by Arab merchants. Last of the four south-eastern rivers is the Han Kiang, which flows within the borders of Kwangtung province and has Swatow at its mouth. All four of the cities mentioned are ports which figured prominently in the 1870's when the China tea trade was in its heyday. At these ports the famous tea clippers gathered during the summer and set out on their race to London.

The climate of this south-eastern area is essentially sub-tropical. The Tropic of Cancer passes right through Swatow Bay. Thus the whole of Fukien and the southern half of Chekiang, which together form the region, lie outside the tropics. The broad belt of mountainous country along the whole north-west border affords protection against the cold north winds in winter. Winter temperatures seldom fall below freezing at sea-level in the south but are considerably lower in the north. Summers everywhere are hot. This is a region of heavy rainfall with a marked summer maximum, usually more than two thirds of the total precipitation occurring in the five summer months, May to September. This, too, is the Chinese coast most affected by typhoons, which may be expected any time between May and October, although their main incidence is in August and September. They originate in the Pacific to the east of the Philippines and Taiwan and usually move west and north-west, many striking the Kwangtung and

Fukien coasts. Typhoons bring with them torrential rains and hurricane winds of 100 m.p.h. and more. Barometric pressure may fall an inch and a half to two inches in a few hours; the sea-level rises alarmingly and with disastrous results to shipping and property bordering the sea. It is fortunate that these typhoons die out quickly as they penetrate inland.

	Jan.	Feb.	Mar.	Apl.	May	Jun.	Jly.	Aug.	Sep.	Oct.	Nov.	Dec.	Total
Wenchow													
Temperature (°F.)	47	47	53	61	69	77	83	83	78	70	60	51	
Rainfall (inches)	1·8	3·4	4·9	5·5	7·0	9·9	7·5	9·4	8·0	3·2	2·1	1·6	64·3
Foochow													
Temperature (°F.)	53	52	56	64	72	80	84	84	76	67	59	50	
Rainfall (inches)	1·8	3·8	4·5	4·8	5·9	8·2	6·3	7·2	8·4	2·0	1·6	1·9	56·4
Swatow													
Temperature (°F.	59	57	62	70	77	82	84	83	82	76	68	62	
Rainfall (inches)	1·4	2·5	3·1	5·6	9·0	10·5	7·8	8·4	5·5	2·9	1·6	1·5	59·8

The many and high ranges of the Wuyi Mountains have served to isolate the south-east from the rest of China to a considerable degree. The original inhabitants, the Min Yueh, remained independent of Chinese rule until nearly the end of the Later Han dynasty. Subsequent pressures from the north have resulted in many infiltrations of people, notably the Hakka or 'Guest Folk' during the ninth and tenth centuries when Mongol (Khitan) pressure was at its height. The mountainous nature of the country with its isolated valleys has resulted in many local and widely differing dialects which virtually constitute a different spoken language from Mandarin, although it must always be remembered that the written language throughout China is the same. The people are ruggedly independent and are reputed to have less polished manners than their compatriots of the north and south.

Geography has influenced livelihood in the south-east to much the same extent and in much the same way as it has done in Norway. Topography is the ruling factor. The mountainous character of the area and the deep, narrow valleys limit farm land to about 8 per cent of the whole.[1] Although, like Kwangtung, the soil is acid and not very fertile, this small proportion is farmed most intensively. Rice, as elsewhere, is the main crop and grown wherever possible. The growing period is long enough to enable two crops per annum to be taken. Other important food crops are sweet potatoes, vegetable oils (rapeseed and sesamum-seed), beans, groundnuts and vegetables. Sub-tropical fruits—citrus, peaches, persimmon—and sugar cane are produced in quantity.

This is the most famous tea growing area of China, especially in the centre and north, around and behind Foochow and Wenchow, whence

---

[1] G. B. Cressey, 'Land Forms of Chekiang', *Annals of A.A.G.*, Vol. 28, 1938.

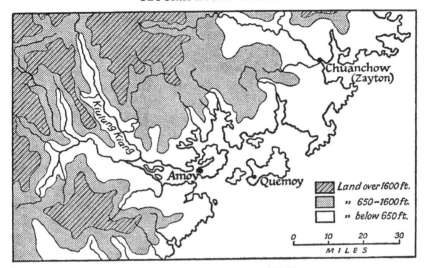

FIG. 65. The rugged coast of Fukien

come the Bohea teas. Every summer during the 1870's and 1880's the tea clippers used to gather at the ports to await the tea harvest. On their outward passage in early summer they sailed south-west on the North-east Trades until the South American coast was reached; then, with the South-east Trades on their port side the clippers hugged the Brazilian coast until the Roaring Forties were encountered. Turning due east the sailing ships used these westerlies until they had nearly reached Australia, when again the South-east Trades were used to carry them north-westward. After the clippers had passed through the Sunda Straits and crossed the Equator, the SW monsoon carried them to the South-east China coast. The return passage in the autumn made similar use of the prevailing winds by following quite a different course. By the time that the tea had been harvested, processed and packed, the SW monsoon had given place to the NE monsoon. These following winds were used right across the Indian Ocean to Madagascar. With the SE Trades on their port the sailing vessels maintained their south-westerly course, and by hugging the South African coast, the Cape was rounded with the least possible encountering of the headwinds of the Roaring Forties, which at this season are much farther south. Thereafter the SE Trades carried the clippers across the South Atlantic to Cape Sao Roque. Rather than turn directly homeward from here and face headwinds from the NE Trades, the clippers continued to the Bahamas with the NE Trades on their starboard side and then turned for home across the North Atlantic on the westerly Anti-trades. The first

263

clipper home with its tea cargo secured not only the best price but gained considerable prestige in shipping circles. An exciting account of one of these races is given by John Masefield in his novel *Bird of Dawning*.

The other resource that this mountainous south-east coast region offers comes from its forests. Heavy rainfall and sub-tropical temperatures ensure the production of a good forest cover of conifers (*Cunninghamia* and pine) camphor and bamboo. The latter provides material for innumerable purposes from pens to scaffold poles, and from chopsticks to tables. Timber cutting has been heavy and there has been some deforestation, especially in the more accessible regions along the lower reaches of the rivers, but it has not reached the serious dimensions of the north and centre.

Restricted agriculture has turned the faces of the people seaward to look for a living. The many good natural harbours and the plentiful timber supplies for shipbuilding have combined to develop here the fisher-folk of China. The thousands of bays and inlets afford good grounds for inshore purse seine fishing and oyster culture. Many of the larger junks, which form the deep-sea fishing fleets operating from Ningpo, Wenchow, Foochow and Amoy, have been mechanized. These ports also supply a large proportion of the merchantile fleet of junks which carries the very considerable coastal trade extending not only as far north as Tientsin but inland up the Yangtze as far as Ichang. As with Devon in England, so Fukien in China has become the region from which its sailors have been recruited. Fukien sailors supply most of the admirable crews which man so many of the western steamship lines.

In spite of the intensive use of its limited agricultural land and the full exploitation of its timber and fishing resources, Fukien has been unable to meet the pressure of its increasing population. This pressure, which was intensified in the last two decades of the nineteenth century by the failure of the tea trade through Indian competition, has been relieved during the last century by emigration mainly to Malaya, Borneo and Indonesia. It is not without reason that the People's Government has provided special investment facilities in Fukien, free from socialization, for remittances sent home by these emigrants.

5. THE YUNNAN-KWEICHOW PLATEAU

This plateau region is a south-western extension of the Tibetan Plateau and is composed generally of rough mountainous land, having a general level of 6-7,000 ft. above sea-level in the west, with many snow-capped peaks rising to over 15,000 ft. The plateau descends eastward, many parts of Kweichow having a general level of less than 3,000 ft., the mountains being less rugged and the valleys wider.

Western Yunnan is characterized by north-south ranges, between which the Mekong, Salween and Irrawaddy flow and which have constituted so great a barrier between China and Burma in the past. Deep, steep-sided and heavily wooded valleys with fast-flowing, unnavigable and often un-fordable rivers make communication in these parts extraordinarily difficult. Many of these valleys are 5,000 ft. deep and perhaps only $\frac{1}{4}$ of a mile wide at that height and within hailing distance. Yet it may demand a whole day's strenuous journey to get from one side to the other. Occasional swaying bamboo rope bridges, slung 500 ft. above the rocky chasm, help the traveller across. It was in these malarial infected valleys that Kublai Khan lost nearly half his forces when invading Burma. Farther south these ranges are lower, less precipitous and more easily crossed. This western region has been heavily folded and is still unstable and subject to earthquakes. A severe one in 1925 resulted in great loss of life and had long-lasting effects on agriculture. The comparatively slight changes in level caused by earthquakes spell disaster to paddy farmers, whose fields must of necessity be absolutely level if they are to be flooded.

The character of the plateau changes eastward. Central and south Kweichow and south Yunnan are composed largely of Devonian limestone and have developed the same fantastic karst landscape which we noted in West Kiangsi. In the western part of the plateau the erosion cycle has not advanced as far as it has in the east and much of the drainage system is underground. A feature of the karst region is the large number of small isolated valleys, known locally as *patze*, which are reminiscent of the *polje* of Yugoslavia. They are deep and often oval in shape, giving them the appearance of arenas, especially as their sides are lined with the irregular terraced fields which follow the contours of the steep valley slopes.

The Kinsha (Upper Yangtze) cuts a deep trench along the north side of the plateau from which it receives many short right-bank tributaries in the west and the much larger Wu Kiang in the east. The latter drains the greater part of northern Kweichow, which is lower and a gentler topography than the rest. Thus the Yangtze drains the northern regions of the plateau; the Mekong, Salween and Irrawaddy drain the west; and the Red River and Si Kiang the south and south-east.

The climate of this plateau is the most equable in the whole of China. Cressey says of it, 'The province of Yunnan is an island, not of land amid an ocean, but of moderate temperature and clear skies, surrounded on three sides by hot, humid lowlands'.[1] In spite of the height of the plateau, winters are mild and summers are cool by comparison with the plains to the east and south. Those who, during the Sino-Japanese War (1939–45), were forced to trek from Central China to these regions, expatiate on the bright, clear winters and the delectable summers of Yunnan, the meaning of which is 'South of the Clouds'—the clouds being those of Szechwan. It should be added that Kweichow does not enjoy the same reputation. It is a province with a high incidence of cloud and a much more even distribution of rainfall than Yunnan. Kunming has a range of only 24°F. between average January and July temperatures. Annual rainfall ranges between 40 and 45 in. for most low-lying places, although slopes exposed to moisture-laden winds have a much heavier fall. Most places are free from frost for at least ten months of the year; there is thus a long growing period.

	Jan.	Feb.	Mar.	Apl.	May	Jun.	Jly.	Aug.	Sep.	Oct.	Nov.	Dec.	Total
Kweiyang													
Temperature (°F.)	37	42	53	63	71	72	76	77	68	57	53	47	
Rainfall (inches)	1·0	1·1	0·9	2·9	7·0	8·1	9·0	4·1	5·4	4·2	2·0	0·6	46·3
Kunming (5,940 ft.)													
Temperature (°F.)	48	51	60	66	70	72	70	70	66	63	56	49	
Rainfall (inches)	0·5	0·5	0·6	0·7	3·8	6·1	9·4	8·2	5·4	3·6	1·7	0·6	41·1

The plateau is well forested with a mixture of conifers (mainly *Cunninghamia*) and broadleafed deciduous and evergreen trees according to altitude. Mountain steppe of tall grasses are also common, and the whole district is renowned throughout China for its wealth of wild flowers—its azaleas and rhododendrons.

The main types of soil are mountain red earths and mountain yellow podzolic earths, which occur nowhere else in China.[2] Neither results in good agricultural soil. The red earths are badly eroded and leached. The yellow earths, developed under mixed forest of deciduous, coniferous and evergreen broadleafed trees in which deciduous predominate, are acid. The only good agricultural soil is the purple-brown forest soil and this, unfortunately, is very limited in extent.

Favourable climate, giving a long growing season, which should lead to heavy agricultural production, is offset by the limited area of cultivable land and the poor soils. Only about 5 per cent of the total area is actually cultivated, although Buck estimated in 1937 that about 63 per cent of the uncultivated land was put to some productive use. Production is confined

[1] G. B. Cressey, *Land of the 500 Million* (New York 1955), p. 227.

[2] Ma Yung-chih, *General Principles of Geographical Distribution of Chinese Soils* (Peking 1956).

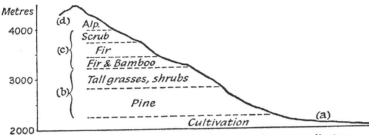

(a) *Alluvium. Some acid paddy soil. Some calcareous alluvium*
(b) *Mountain red and yellow earth*
(c) *Mountain podzolic soil*
(d) *Mountain stony soil*

FIG. 66. Profile of Yunnan vegetation

to the *patze*, where rice is the main summer crop. In south Yunnan there is a little double cropping of rice. Other summer crops are corn, barley and millet, sweet potatoes, tobacco, vegetables and some tea of high quality. Corn and sweet potatoes have formed the main staple food of the poorer peasants. The main winter crops today are oil seeds, beans and vegetables. In the past a great deal of opium was grown on account of its high cash value and the ease of its transport.[1] This was harvested just before the planting of the rice in spring and afforded an income larger than any other winter crop. Its cultivation is now forbidden.

More than half the population of the plateau is composed of the many aboriginal peoples and tribes who, in the course of preceding centuries, have been forced back into the mountains by the Chinese as they pressed south and west, rather as the Britons were forced back into the Welsh highlands by Saxon and Dane. These mountain folk tend to keep to their own modes, customs, dialects and dress. Until the 1949 Revolution they carried on trade with the Chinese of the lower lands. Usually they specialized in and marketed only one product, e.g. the Black Lolo sold fuel; the White Lolo, corn; the Miao-tzu, corn; the Pei I, fruit.[2] Since 1949 most of these people have come under local administration, known as Autonomous Chou or Districts designed, at least in part, to prevent their exploitation.

The Chinese or Han part of the population forms the farming community of the lower lands and is engaged mainly in rice cultivation. This has been one of the most conservative and reactionary areas of China—easy going

[1] Fei Hsiao-tung and Chang Chih-i, *Earthbound China: a study of Rural Economy in Yunnan* (Chicago 1945).

[2] Lossing Buck, op. cit.

and lacking in energy and initiative. Before the land reform measures of 1949 there was considerable absentee landlordism. The main objective of owner and tenant alike appears to have been to find contentment, and by contentment was meant the avoidance of painful labour as far as possible. This led to much exploitation of women as labourers; a daughter-in-law was thought of in terms of a cheap, long-term labourer.[1] The lack of energy in the past can be attributed to the very widespread habit of opium smoking.

There has always been hostility and suspicion between the indigenous tribes and the settled Chinese. This has resulted in unrest and turbulence, which has been further accentuated by the presence of a body of some half a million Moslems who have come south from Kansu. In 1873 a great Moslem rebellion broke out and it was not finally quelled until the nineties.

The conservatism referred to above is due in no small measure to the isolation resulting from poor communications. The rugged topography makes all travel and transport extremely difficult. Until recently practically all transport was by way of coolie pole along paths at best paved with slabs of stone. There was an old trade route, which ran from Suifu (Ipin) south to Yunnan-fu (Kunming). Here it divided. One branch ran westward to Tali and Tengyueh and so to Bhamo in Burma. The other turned south to Mengtze and Tongking. It was along this latter route that the first railway penetration into Yunnan was made. The railway was built with French capital and its completion in 1908 was no small engineering feat, especially in the Chinese section where literally hundreds of tunnels were bored and thousands of bridges and culverts had to be built. Such was the state of the mountain paths in the north that it was quicker and cheaper to travel from Chungking to Kunming by going down the Yangtze to Shanghai,

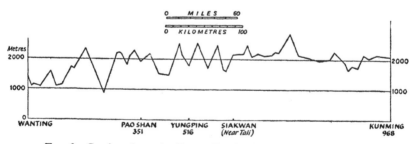

FIG. 67. Section along the Burma Road from Wanting to Kunming

[1] Fei Hsiao-tung and Chang Chih-i, op. cit.

268

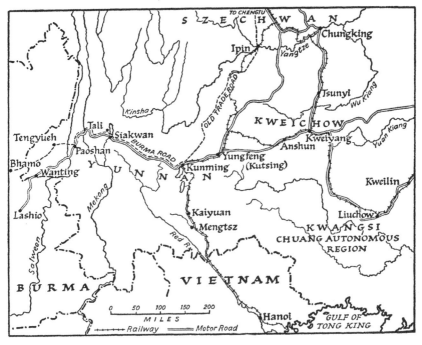

FIG. 68. Yunnan-Kweichow

thence to Hanoi and so by rail to Kunming than by taking the overland route.

The influence of the Hanoi-Kunming railway in breaking down the isolation of the plateau was small as compared with that of the Second World War, when the Japanese invasion of the Yangtze and Si Kiang Basins drove millions of Chinese of all classes into the Szechwan Basin and up on to the Kweichow and Yunnan plateau. Schools, colleges, universities, industrialists, manufacturers and merchants alike moved in, bringing with them ideas and techniques which inevitably have shaken the old conservatism. The war also led to the construction of the Burma Road from Lashio via Kunming to Chungking and Chengtu.

Since 1949, under the first two Five Year Plans, railways were projected to connect Chengtu with Kunming via Suifu (Ipin) and Yungfeng, and between Chungking and Kunming via Kweiyang and Yungfeng. This network is now nearing completion and will help considerably in opening up the region.

The Yunnan-Kweichow plateau is not richly endowed with coal or iron. Coalfields, adequate for regional development of light industry, are found near Kunming, at Anshun, west of Kweiyang and at Kaiyuan to the north of Mengtze. Its wealth lies in its deposits of non-ferrous minerals of tin, copper, zinc and phosphorous. Before the Second World War Yunnan produced between 6 and 7 per cent of the world's tin. Copper has been worked for many centuries. It has been suggested that the bronze (tin and copper) of the Yin or Shang dynasty came from here. Tali marble is sought after throughout China for decorative building material on account of its striking and beautiful markings.

6. THE NORTH-EAST (MANCHURIA)

The North-East, or Manchuria as it is more generally known to Westerners, has been associated with China Proper since very early times in varying degrees of intimacy. Under the Earlier and Later Han (206 BC–AD 220) the lower Liao River region, the East Manchurian Uplands and North Korea owned Chinese sovereignty but reverted to nomadic tribal leadership on the fall of the dynasty. In T'ang times (AD 618–906) a large area, reaching well to the north of the Liao Basin, formed a kind of Chinese protectorate, while under the Chin Empire (AD 1125–1206) Chinese territory extended as far as the Amur River and had a long seacoast on the Sea of Japan. This, and a great deal more, was incorporated in the great Mongol Yuan Empire under Kublai Khan and remained under the nominal sovereignty to China under the early Mings in 1415.

Far-flung though these periodic extensions of territory may have been, that there was no real expansion to the north-east intended—or to the north-west for that matter—is demonstrated by the fact that part of the Great Wall was built here and the purpose of this wall was to bar entry and to some extent to prevent exit. The wall was extended in a wide semi-circle to enclose the lower Liao and it thus formed a kind of Chinese Pale. Thus the coast route via Shanhaikwan into China, which was used for trade, was covered but the more difficult upper Liao route, which was more usually used by invading forces, was not defended. Owen Lattimore, at pains to emphasize the restrictive nature of the Wall, says 'A positive expansion does not build limiting walls. There are no Great Wall systems in the South'.[1]

Profiting by internal disorders due to civil war and endless court intrigues

---

[1] Owen Lattimore, 'Chinese Colonization in Manchuria', *Geographical Review*, 1932.

of a decadent Ming dynasty, Nurhachu, a Manchurian nomad chief, first conquered the surrounding East Mongolian tribes and gained their support, rather as Jenghis Khan had done before him. He then proceeded to build up a professional army of eight corps or 'banners', which owed no nomadic tribal or territorial allegiance and which were prepared to serve anywhere with a promise of a share in the fruits of victory. With this army Nurhachu overran the Chinese Pale in 1626. Although he died in that year the invasion continued and Peking fell in 1644. The militarily powerful but administratively inexperienced Manchus made peace with, and secured the cooperation of, the Chinese gentry, who had been direly persecuted in the last years of the Mings, and were thus able to establish governmental control over the whole country.[1]

The first great Ch'ing dynasty emperor, K'ang-hsi, embarked on a policy of keeping the North-East as a closed preserve for Manchurian bannermen, who became a privileged class both here and throughout China, receiving a subsidy from the wealth of China. In 1668 K'ang-hsi closed Manchuria to the Chinese, although, up to 1644, the country was underpopulated. The Manchurians are reported to have captured and brought to the North-East more than one million Chinese during the many sporadic raids they made on China during the declining years of the Ming Empire.[2] A willow palisade and moat were constructed encircling and isolating the Manchurian preserve, but the prohibition was never very strictly observed. On the one hand, whenever famine occurred in north China, especially in Shantung, there was an influx of refugees into Manchuria which would have been very difficult to control. On the other hand, the bannermen, never great agriculturalists, became demoralized through being subsidized and came to rely more and more on imported Chinese labour. By 1779 it is estimated that there were more than 6 million Chinese peasants working in each of the provinces of Kirin and Fengtien (modern Liaotung). The need of the great landowners was for workers rather than tenants. Thus began the seasonal migration of labour, mainly from Shantung, which was maintained up to the outbreak of the Second World War. As might be expected there was always a residue which remained behind, but generally speaking the objective of the migrant was to make his money and to return to his ancestral home.

During the latter half of the nineteenth century Russia's penetration eastward in Siberia and along the banks of the Amur River caused grave

---

[1] W. Eberhard, *A History of China*, Chapter 12 (London 1948).

[2] Ho Ping-ti, *Studies on the Population of China* (Cambridge, Massachusetts 1959).

concern to the Ch'ing government. The wide open spaces of Heilungkiang, which is the Chinese name for the Amur River and means 'the Black Dragon River', needed populating. In order to encourage colonists, the whole of this northern region was thrown open to Chinese immigrants. It is reported that more than 100,000 households had moved in by 1870. Part of this colonization in the extreme north, however, was by 'garrisons' of yeoman stock having some military tradition, whose services could be called on if necessary.[1]

In 1907 all legal obstacles to immigration were removed. Between 1920 and 1930, owing to revolution, civil war and widespread famine in north China, there was a continual and increasing migration into Manchuria. This took the form mainly of seasonal labour movement, but every year there was a residuum of between 40 and 60 per cent which did not return home. This colonization was of the refugee type. The colonist left his home with regret and with an eye to eventual return in contrast to the backwoodsman of North America of the last century, who set out in high adventure to carve a new world for himself. Immigrants before 1931 were mainly agriculturalists. Thereafter their composition changed, until by 1938 they were mainly industrial workers and miners.

Volume of immigration and emigration from Manchuria[2]

Year	Volume of Immigration	Volume of Emigration	Residuum	% residuum to Immigration
1923	342,038	240,565	101,473	30
1924	376,613	200,046	176,567	47
1925	491,948	239,433	252,512	51
1926	572,648	323,566	249,082	43
1927	1,016,723	338,082	678,641	67
1928	938,472	394,247	544,225	58
1929	1,046,291	621,897	424,394	41

Population rose very rapidly. In 1904 Sir Alexander Hosie estimated that the population was about 17 millions. The Research Bureau of the South Manchurian Railway Co. placed it at 34 millions in 1930 and the census figure for 1953 was 47 millions. Since 1953 colonization, particularly in the northern regions of Heilungkiang and Kirin, has continued at great

[1] O. Lattimore, op. cit.

[2] Franklin L. Ho, 'Population Movement to the North-eastern Frontier of China', quoted from Ho Ping-ti, op. cit.

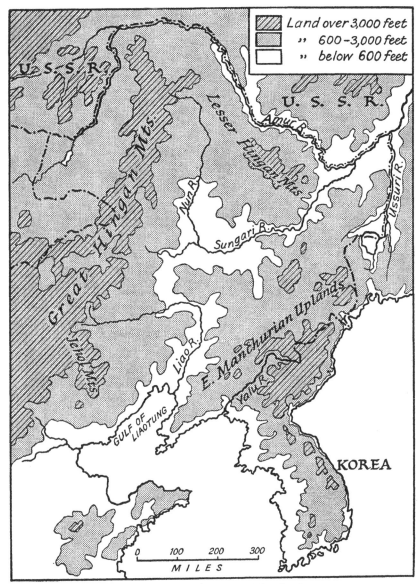

FIG. 69. The North-East: Manchuria

speed and would appear to have been much more akin to the nineteenth century American pattern in its enthusiasm and optimism, but unlike it in that
the lands settled were held corporately and not individually. The rolling
plains of the Sungari and the Nonni have come under the plough; the slopes

273

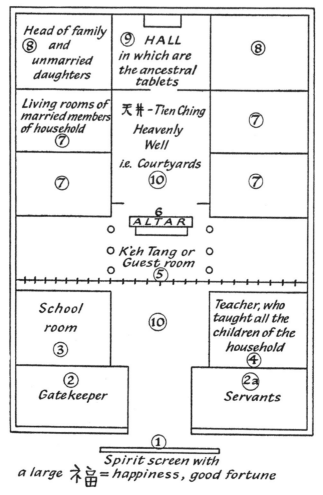

FIG. 70. Plan of an old style house of a well-to-do Chinese family

of the Ta Hingan (Great Khingan) under the forester, lumberman, geologist and prospector.

George B. Cressey, in his *China's Geographic Foundations* makes a felicitous comparison of the Manchurian plains to the courtyard of a Chinese house. Above is a plan of a house of a fairly well-to-do Chinese family of pre-1949 times. It will be seen how much of it reflects the physique of the country. True, there is nothing to equate with the spirit screen, nor is there any back door to match the outlet of the Sungari at its confluence with the Amur. The Liao River Basin and the Sungari correspond clearly enough with the courtyards, and the waterparting between them matches the *K'eh t'ang* or guest room. The Jehol Mountains as they come down steeply to the west shore of the Gulf of Liatung at Shanhaikwan correspond to the gatekeeper's lodge and the ancient Hills of the Liaotung peninsula on the east with the servants' quarters. The Ta Hingan (Great Khingan) Mountains with their steep scarp faces to the east and gentler dip slopes away to the Gobi, correspond to the living rooms on one side and the East Manchurian Uplands to the living rooms on the other. The Siao Hingan (Little Khingan) Mountains running east and west in the north match the hall, and so complete the picture.

The plains of the Liao and the Sungari Rivers are largely erosional in contrast to the North China Plain, which is depositional. The Liao River is shallow and carries a heavy load of silt, which is encroaching on the Gulf of Liatung at a very rapid rate. Old Newchuang, which was originally a port at the mouth of the river, today stands miles from the coast and now has Yingkow as its port. This silting of the Gulf makes the siting of a port there difficult and accounts in part for the importance and development of Dairen and Port Arthur on the eastern side of the peninsula. In consequence of its shallowness the Liao is of little use for navigation save for light junk traffic. In contrast, the Sungari is larger and deeper and is navigable for river craft of 1,000 tons and more, which carry an important amount of both passenger and cargo traffic. The Sungari empties into the Amur or Heilungkiang, which marks the entire northern border of Manchuria and which has even greater navigability than the Sungari. With the recent agricultural and industrial development of this region both in the USSR and in Northern Manchuria the importance of these two rivers has been greatly enhanced.

While the influence of the monsoon rhythm is still evidenced in the North-East by the seasonal distribution of rainfall, its continentality is strongly emphasized by the big range of its temperatures which increase markedly from south to north. Clear winter skies and strong, bitterly cold winds from the Siberian high produce minimum temperatures at times as low as $-30°F$. at Mukden (Shenyang) and $-40°F$. at Harbin. Winters are long. Harbin has five months with average temperatures well below freezing. The spring and autumn seasons are short as the rapid rise and fall of temperatures in April and November respectively indicate. Summer temperatures on the plains average between 70 and 75°F. With the rise in temperature comes the rain, about four fifths of which fall in the four summer months. The amount of rainfall decreases from south to north and from east to west. It is heaviest in the East Manchurian Uplands

(Changpai Shan). Nearly everywhere on the plains it is adequate for wheat growing, i.e. more than 15 in.

Harbin (525 ft.)	Jan.	Feb.	Mar.	Apl.	May	Jun.	Jly.	Aug.	Sep.	Oct.	Nov.	Dec.	Total
Temperature (°F.)	−2	5	24	42	56	66	72	69	58	40	21	3	
Rainfall (inches)	0·2	0·2	0·4	0·9	1·7	4·1	5·8	4·2	2·2	1·2	0·4	0·2	21·5
Mukden (144 ft.)													
Temperature (°F.)	9	14	28	46	60	70	76	74	62	48	29	14	
Rainfall (inches)	0·2	0·2	0·8	1·1	2·2	3·4	6·3	6·1	3·3	1·6	1·1	0·2	26·5
Dairen (315 ft.)													
Temperature (°F.)	23	26	36	49	60	69	75	77	68	57	41	28	
Rainfall (inches)	0·5	0·3	0·7	0·9	1·7	1·8	6·4	5·1	4·0	1·1	1·0	0·5	24·0

The natural vegetation of the plains is that of temperate zone grasslands. This, however, is rapidly disappearing as more and more of the plains are passing from nomadic pasturage to ploughland. The resultant soil is chernozem in character but is not considered light enough by some authorities to be placed firmly in that category. The mountain lands of the Ta Hingan (Great Khingan) Jehol Mountains and the East Manchurian Uplands are forested. As might be expected, the latter, receiving the heaviest rainfall, are the most densely wooded with pine, larch and fir. The massive Manchurian pine is the best timber and has been cut, during the last fifty years and particularly by the Japanese during the Second World War, to an alarming extent. The eastern facing slopes of the Ta Hingan Mountains are fairly well covered, but the forests quickly fade out on crossing their crests and descending the western dip slopes. Forest gives place to steppe and eventually to semi-desert and true desert. There are far-reaching and long term plans to create a wide, continuous forest belt along the whole length of the eastern side of the Ta Hingan and to conserve more carefully the valuable timber of the East Manchurian Uplands. In 1945 it was estimated that standing timber covered about 54 million acres, whilst 163 million acres were in urgent need of re-afforestation.

Modern development of Manchuria stems out of the economic and political ambitions and rivalries of the Great Powers, most notably Russia and Japan, and is intimately related to railway construction. We have seen something of the earlier breakdown of Chinese isolationism mainly by the British earlier in the nineteenth century and the relative quiet which followed the Second Opium War in 1860. This quiescent period was broken first by the outbreak of the Sino-Japanese War (1894–5) which synchronized with growing Russian activity in eastern Siberia where she was building the Trans-Siberian Railways. This railway, if built entirely on Russian soil, would have had to follow the long loop of the Amur-Ussuri valleys to reach Vladivostok. Accordingly Russia sought and obtained a concession in 1896 to build the line across Manchuria via Tsitsihar and

Harbin to Vladivostok. This agreement included not only the right to build the railway, which was called the Chinese Eastern Railway, but also the right to develop the land alongside the line and the granting of some measure of extraterritoriality to Russian subjects. This concession was further extended to permission to build the line from Harbin to Port Arthur thus giving Russia an ice-free port on the Pacific.

So broke out again the scramble among the Great Powers for zones of influence and concessions in a decadent and impotent China. Germany gained Tsingtao and its environs in 1897; Britain obtained Weihaiwei. In addition to the annexation of Korea and Formosa, Japan claimed the Liaotung peninsula as the fruit of her victory over China, but this was denied her through the strong opposition of Russia, backed by France and Great Britain. Mutual fears and fierce rivalry between Russia and Japan broke into open war in 1904, when Russia was signally defeated by Japan to the surprise of the rest of the world. As a result Japanese influence surplanted that of Russia in Liaoning. The development of Manchuria from that time on steadily became an integral part of Japanese imperialist policy for the creation of a Greater Japanese Empire, which culminated in the Sino-Japanese War (1937–45) and Japan's entry into the Second World War.

Until 1931 Manchuria remained nominally subject to the Peking Gov ern-ment, whether Manchu or subsequently Republican, although its writ seldom ran, but the direction of its economic development fell more and more into Japanese hands. In 1931 Japan, flaunting world condemnation through the League of Nations, virtually annexed Manchuria, forming it into the State of Manchukuo and making the ex-Manchu emperor, deposed in 1911, its puppet head. Thereafter Japanese dominance remained complete until after her defeat at the end of the war in 1945. Then followed a race and a struggle between the Nationalist Kuomintang and the Communist Kungch'antang for mastery, won eventually by the Communists.

It is with this history in mind and in this political setting that the economic geography of the North-East must be examined, for Japan regarded the country purely as a means of supplying sinews for her power whether in the provision of foodstuffs for her home population, raw materials for her industries, or an outlet for her swarming people. To this end Japan lavished millions of pounds of capital on Manchuria in developing its many resources.

Some idea of agricultural development can be gleaned from the increase of land under cultivation. In 1915 some 16·5 million acres were being tilled. In 1933 this had doubled at 33·3 million acres and in 1939 it had grown to

43·7 million. During the Second World War there was probably little agricultural development due to concentration on the production of armaments and the consequent shortage of labour and the curtailment of farming machinery. However, there is every reason to believe that since 1946 agricultural development has gone forward at an even greater rate. Farming in Manchuria has been of a more extensive nature and the farms have been much larger than in China Proper. Consequently the region has lent itself more readily, after Land Reform in 1949, to collectivization. Collective farms, state farms and large communes are the normal organization today.

The growing season is short, especially in the north, and only one crop can be secured each year. The two outstanding crops are kaoliang and soy-beans, the character and uses of which have already been described.[1] The great demand for soy-bean oil in the First World War for the manufacture of explosives sent production soaring, and until the outbreak of the Second World War Manchuria was the main world supplier. During that war and ever since, the USA has become an important grower and Manchuria's share in world trade has fallen, partly for this reason and partly because China is absorbing more in its own growing industries. Wheat is the grain grown by the settlers, who push north into the new lands of the Sungari and Nun (Nonni) plains, whilst corn is a more popular crop in the south. Millet is widely grown. Other crops are groundnuts, sweet potatoes and Irish potatoes. Sugar beet was introduced by the Russians in 1909 and today it is an important crop, together with cotton and hemp. Some rice, mainly upland, is grown. Opium used to be a popular cash crop but it is now grown only under licence.

In 1904, when Japan took over the Russian sphere of influence in Manchuria, she formed the South Manchurian Railway Company, which was a body entrusted not only with railway building but also with the development of public utilities, industries and ports, which the railway was designed and intended to serve. By 1931, when Japan virtually annexed the country, a railway network of nearly 3,750 miles (6,000 km.) had been constructed and the line from Harbin to Dairen had been double tracked. By 1943 this figure had risen to 9,375 miles (15,000 km.), which was more than the rest of China's railways put together. Up to 1945 it is estimated that the South Manchurian Railway Company had invested £350 million. Thus it was that, when the present government gained power in the region, it was not faced with the dilemma of whether to concentrate first on the

[1] See pp. 122 and 130.

development of communications or of industry: the essential network was already there.

The natural resources awaiting development were vast and widespread. Coal reserves, while not equal to those of the Shensi-Shansi field in either quality or quantity, are yet enormous, and what is more, are varied and more accessible than those of the west. The greatest reserves are sited in Liaoning at the two great open cuts at Fusin (Fushin) and Fushun, comprising 4,000 million tons and 950 million tons respectively. Both are of good bituminous steam coal and have phenomenally thick seams. At Fushun the coal seam varies between 120 and 360 ft. in thickness and averages 240 ft. It is a 'hard, smoking and asphalt' coal with the best quality at the base of the seam. The coal lies below a stratum of oil shale 240 ft. thick, which in its turn is covered by a layer of green shale of equal thickness. These, together with the coal, are exposed along a cut 1,000 to 1,500 yd. in width for some 4 miles and can be extended for a further 4½ miles in a north-easterly direction. The seams are mainly horizontal, dipping from SE to NW.

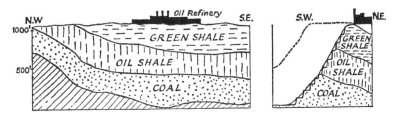

FIG. 71. Sections of Fushun coal cut

The presence of coal here was known 800 years ago, but there was no large-scale working of the cut until 1914, when it began to be worked under Japanese management and capital. In 1944 about 10 million tons of coal were produced, using largely modern mechanized methods of grabs, drills, trains and funicular. After 1945 production fell: in 1949 only 2½ million tons were raised. Then followed a period of rehabilitation. By 1955 the Japanese peak production of 1944 had been nearly regained and now it is claimed that double this amount is produced annually. The Fusin (Fuhsin) cut is reported to have made similar advances in production.

In order to use the 600 ft. or so of overburden of oil shale and green shale, the Japanese built an oil refinery and distillery close to the deposits. This plant was badly neglected during the period 1945–9, i.e. between the fall of Japan and the rise of the present régime. By 1955 it had regained its former output. The Japanese experimented in the use of the green shale for soap manufacture, and it is now being used for water softening. The removal of the overburden presents an increasing problem. Not only does the amount to be moved increase as the strata dip, but sooner or later it will necessitate the removal of the oil refinery, which covers a large area and which now stands near the edge of the cut.

Associated with the coal and oil industry in this region are some of the most advanced workers' welfare and health services in modern China, including public baths, libraries, clubs, hospitals, clinics and free medical services, workers' insurance and old people's homes.

Another large coalfield in Liaoning is Hsinan to the north of Shenyang (Mukden), which has an estimated reserve of 600 million tons of good bituminous coal. Peipiao, south-west of Fuhsin, Tungpientao, near the Korean border and Penki (Penhsihu), south-west of Shenyang all produce bituminous coking coal and have reserves of 250, 236, and 220 million tons respectively. The best coking coal, however, comes from Haolichen (Haolikang) in the north-east, where there is a reserve of some 600 million tons. Also in the north is the bituminous-lignite field comprising 300 million tons at Mishan.

To match this wealth of coal there is a fair amount of iron ore, but unfortunately most of it is of lean low-grade (33–35 per cent). Japanese estimates of reserves in 1945 were 4,459 million tons of low-grade and 59 million tons of high-grade (50–63 per cent), mainly haematite, the main fields being at Anshan, Penki (Penhsihu) and Tungpientao. Most of the non-ferrous minerals necessary in the production of iron and steel (limestone, dolomite and fire-clay) are found in adequate quantities in South Manchuria. There are also considerable reserves of molybdenum, copper, lead, zinc, graphite and bauxite.[1]

During their twelve years of occupation of the North-East the Japanese were assiduous in the development of the iron and steel industry. In 1932 there was a plant capacity for the production of pig-iron of 637,500 metric tons, although only 368,181 tons were produced. No steel was made. As a result of the work done during the Five Year Plan, 1936, known as the Manchurian Industrial Development Plan, production was at a peak in 1943 when, with a plant capacity of just over 2½ million metric tons, there was an output of 1,726,700 tons of pig-iron and 837,000 tons of steel.[2]

[1] A. J. Grazdanzev, 'Manchuria, 1945: an Industrial Survey', *Pacific Affairs*, Vol. 18, December 1945.

[2] T. T. Read, 'Economic-Geographic Aspects of China's Iron Industry', *Geographical Review*, Vol. 33, 1943.
A. Rogers, 'Manchurian Iron and Steel Industry', *Geographical Review*, 1948, p. 41.

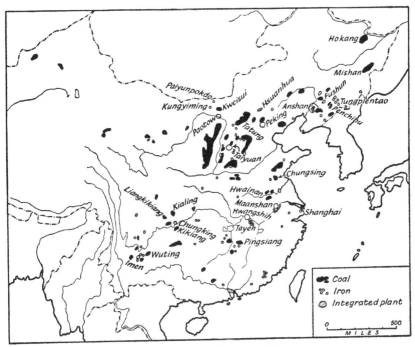

FIG. 72. Coal, iron and integrated industry

The main iron and steel centre is at Anshan, south of Shenyang (Mukden) and on the main line from that city to Dairen. Here, within a radius of 80 miles, all the necessary ingredients for the industry are found: coal from Fushun and Penki (Penhsihu), reasonably good rich local iron ores and limestone from Huolienchai, only 14 miles away. There were 9 blast furnaces, 59 per cent of the pig-iron from which was used for steel making in Anshan itself and 41 per cent sent to Japan for use in industry there. Two other centres of the iron and steel industry of less importance than Anshan are at Penhsihu and at Erhtaochiang, sited on the Tungpientao coalfield. Output in 1943 at Anshan was 3,125,000 tons; at Penki 1,125,000 tons and at Erhtaochiang 849,933 tons.

Not till 1949 could the work of rehabilitation begin. Special emphasis and importance were laid on reconstruction in Anshan, which became the king-pin of the country's initial development as a modern industrial state. The pressing problem of skilled workers was solved partly by the help of over 2,000 Russian experts and technicians, and also by bringing in 20,000 Chinese technicians and skilled men from all over the country. With the aid of Soviet experts eight blast furnaces with a capacity of 3 million tons

281

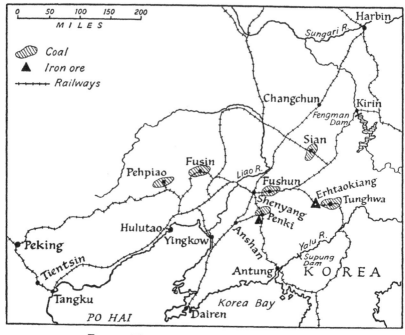

FIG. 73. The industrial heart of the North-East

per annum were in operation in 1955 and a further two were under construction. Heavy rolling mills, covering an area of 60,000 sq. yd. were completed in 16 months and had commenced the production of heavy 40 ft. rails (80 lb. per foot), heavy structural steel girders and angle iron and steel bars for tubes. The design and the plant all came from the USSR and the technicians were trained in the Soviet Union, but in the subsequent extension of these works Chinese designs and Chinese experts were used. The processes, the heating of bars, rolling and shaping, cooling, storing and distributing, are almost entirely automatic. Alongside the rolling mills a seamless steel tubing mill was constructed at the same rapid rate. Here the main production is of tubes of different size and grade for use in boilers, drills, ball-bearings and water pumps. As with the rolling mills, the processes of cutting the bars, heating, boring, cooling, smoothing and finishing are to all intents and purposes entirely automatic.

In 1958 Anshan's steel output was 3·22 million tons and in 1960 over 4 million. Anshan thus became the chief supplier of structural steel for the rest of the country in its attempt to become industrialized in a couple of decades. No. 1 Tractor Plant at Loyang, the Yangtze Bridge, the new Iron and Steel Works at Paotow and at Hankow, the pipelines from Yumen and

Karamai, the heavy and light machine tool shops of Shenyang have all relied on the output of the Anshan works, which compare in size, modernity and efficiency with Port Talbot, South Wales. Anshan, like other places which have suffered a comparable amount of destruction and pillage, enjoys [sic] the advantage of starting afresh with all the most up-to-date equipment.

The hub of the whole industry of the North-East is Shenyang, the modern name for Mukden. It is the capital of the province of Liaoning and is the centre of great engineering activity. A great variety of machinery and tools is produced: three types of standardized lathes are mass-produced in great quantity and distributed throughout the country.

Harbin is the main centre for the production of agricultural equipment. Here the No. 1 Machine Building Plant has three main functions: the production of heavy machinery such as turbines, generators, tractors, combine harvesters, power pumps, etc.; the manufacture of smaller farm implements such as fodder-cutters, winnowers, ploughs, harrows, etc.; and the repair of agricultural machinery and the training, within the repair shops, of mechanics for work in rural centres.

The production of chemicals has been concentrated in Kirin. The Kirin Chemical Co. is a big, integrated enterprise having three big modern plants, which deal with three main products: chemical fertilizers, insecticides, etc.; calcium carbide; dyestuffs. In and around Kirin there are many small workshops which use and re-work the waste products from big industry and which also utilize products from the surrounding forests for making turpentine, resin, pine-wood oil, etc.

Although Manchuria's water power cannot match its great wealth of power from coal, it is fairly well served. The seasonal distribution of rain and the fact that all its rivers freeze in winter are handicaps which can only be overcome by the creation of large enough reservoirs to give an even flow. A Japanese survey in 1940 estimated that, given well sited plants, a maximum of 6 million kW. could be generated. If the great frontier rivers were utilized, there was a further potential of 3 million kW. awaiting development to the mutual benefit of the USSR and China.

The largest development of hydro-electric power carried out by the Japanese was the building of the Fengman dam, 20 miles above Kirin on the Sungari. This huge dam holds back a reservoir of 185 sq. miles with a capacity of 353,000 million cu. ft. and with a maximum generating capacity of 850,000 kW. It was badly damaged and leaking at the end of the war. Repairs were completed in 1957 and new generators installed with a present capacity of 567,000 kW. It should be noted that China is now able to survey, design and build her own thermal and hydro-electric power plants.

FIG. 74. The North-East: hydro-electric power

A second hydro-electric plant has been built on the Mutan Kiang, a right bank tributary of the Sungari at Tao Shan. It has a generating capacity of 383,000 kW. Being built under the Second Five Year Plan is a dam at Kumotsin, above Tsitsihar on the Nun (Nonni) River. This will produce 209,000 kW.

The Japanese planned an ambitious development of the Yalu River on the frontier between Korea and Manchuria, having as an ultimate output 1,133,000 kW. This was never fulfilled. Nevertheless the Supung dam alone has an output of 600,000 kW. Along this valley and associated with the Tungpientao coalfield a considerable industrial centre has sprung up.

7. SINKIANG

As a result of a great campaign by Tso Tsung-tang, Chinese Turkestan was added to the Manchu domains in 1877. It was so called by Westerners to distinguish it from the Turkestan of the Aral-Caspian region, embracing Bokhara, Khiva and Samarkand. Its Chinese name is Sinkiang, meaning New Dominion; under the People's Government its official title is now Sinkiang Uighur Autonomous Region.

Sinkiang contains within its borders two vast basins, the Tarim and Dzungaria and a number of smaller basins, such as the Turfan and Ili, all of which, while differing widely in many respects, have one feature in common, namely, that they are all basins of inland drainage. Only one out of the many rivers finds an outlet to the sea. That one is the Irtysh River, whose headwaters rise in the Altai Mountains and flow for a short distance in Chinese territory before entering their long course across Siberia to the Arctic Ocean.

7a. *The Tarim Basin*

More than half of Sinkiang is occupied by the Tarim River Basin, a spearhead shaped depression, 800 miles long and 400 miles broad at its widest point. Its general level is some 2,000 ft. above the sea in the east, rising to over 6,000 ft. at the western end; it is bounded on all sides by lofty mountains. In the west are the Pamirs, a knot of mountains attaining 18,000 ft. and more, containing high, wide valleys of glacial origin, affording in places good pasture to nomadic tribes but too high for cultivation. Through these mountains are two passes which carry routes from the upper Kashgar River, one north-westward into the Syr Darya to Fergana and Tashkent and one south-west to the Amu Darya valley and to Bokhara and Merv, all names to conjure with in earlier trading days.

The Kunlun ranges, and behind them the Karakoram, form a mighty and forbidding wall on the south-west. Their northern slopes are loess-covered, steep and almost unbroken in spite of their severe erosion. Their ranges rise to 18–20,000 ft. and are most difficult to cross. Routes run from Yarkand to Afghanistan and to Kashmir and from Khotan to Lhasa. The southern border is comprised of the Altyn Tagh, less forbidding, more broken and lower than the Kunlun but still a formidable wall, forming the northern edge of the Tibetan Plateau. From the Kunlun and Altyn Tagh innumerable small rivers and streams, whose sources lie in the snow- and ice-fields of the plateau, pour down on to the plain below.

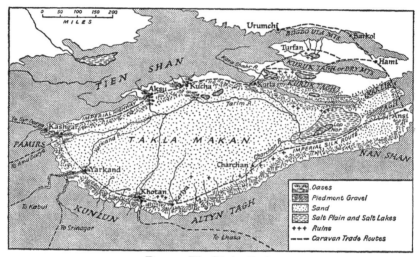

FIG. 75. The Tarim Basin

Running east and north-east from the Pamirs are the many folds of the T'ien Shan or Heavenly Mountains. These form the northern boundary of the Tarim Basin and divide it from both the Ili Basin and the region of Dzungaria. They are neither as high (12,000 ft.) nor as impenetrable as the mountain ranges of the south. The lower but still difficult Karlik (Qarliq) Tagh mark the eastern limits of the basin, beyond which routes pass into Inner Mongolia.

The one long river, from which the basin derives its name, is the Tarim. It has many headwaters in the Pamirs and Kunlun which join to form the Yarkand River. This later loses its name to become the Tarim. It has only one right bank tributary, the Khotan River. Of the many streams flowing down from the southern mountains this is the only one which does not become lost in the sands of the Taklamakan, which they all have to cross, for the Tarim itself hugs the northern borders of the basin. From the southern slopes of the T'ien Shan it receives a number of comparatively large left bank tributaries, fed by the melting snows of the mountains.

The Tarim Basin is comprised of a series of concentric belts. In the heart of the region is the Taklamakan desert, a howling wilderness, true desert for the most part utterly devoid of life and vegetation, a place of desolation of sand and rock.

Ringing this desert is a belt of piedmont gravel of varying width and thickness. This belt has been built up by the detritus brought down by the fast-running streams, which fan out on reaching the plain. It is wider

286

and more marked on the south than the north. The streams disappear under this gravel to reappear and enter the desert in the south. These are the points at which oases are sited along the eastern end of the southern rim of the desert, notably at Keriya, Niya and Charchan (Cherchen). Here there are considerable areas of reeds, and tamarisk and wild fowl abound. In the north and west the large oases, such as Aksu, Kucha, Korla (Kurla), Yarkand and Kashgar, are found above the point of entry into the gravel.

Ellsworth Huntington and Sir Aurel Stein both comment at length on the fact that, far out into the desert, some 70 to 80 miles lower than the present limit of oasis settlement, ruins of large villages and towns, dating between AD 300 and 1,200 are to be found. The remains of dead vegetation indicate a river flow three to four times greater at that time than today.

Above the piedmont gravel belt the Kunlun and Astin (Altyn) Tagh rise steeply. Their slopes are often marked by faults and loess-covered moraines which are barren of vegetation. The rivers descend in impassable gorges which have restricted communication between the people of the oases below and the nomad tribes which occupy the wide valleys of the well-watered plains high up on the plateau, where rich grasses grow. Above these again are higher ranges where temperature restricts vegetation to alpine tundra. Higher still are the ice-fields and névé, the source of all the rivers.

The Tarim, after being joined by the Kara Shahr River on the left bank and the Charchan (Cherchen) on the right, empties into the Lop Nor, a salt lake surrounded by salt marshes and a salt-encrusted plain. This salt plain is bounded on the north by the Kuruk Tagh or Dry Mountains, on the east by the Karlik Tagh and on the south by the western extension of

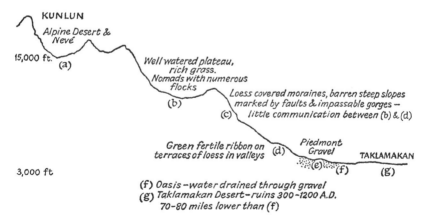

FIG. 76. Sinkiang: section from the Kunlun to the Taklamakan

the Nan Shan. The piedmont gravel is so thickly deposited in the Kuruk Tagh as nearly to obliterate the crests of the ranges.

The oases which ring the Taklamakan are the links in the chain which formed the ancient imperial highways of the Han and later dynasties. These were fixed trade routes which went from oasis to oasis divided from one another by true desert. The Imperial Highway came from the Wei valley to Ansi via Lanchow and Yumen. At Ansi it divided, one branch going north to Hami and Barkol, thence turning west and following the southern foot of the T'ien Shan, passing through the large oases of Korla, Kucha and Aksu to arrive at Kashgar. The other branch, the Imperial Silk Route, passed south of the Lop Nor, through the large oases of Charchan, Khotan and Yarkand and many smaller ones to Kashgar.

It was along this southern route that Huntington gathered his early data in support of this theory of climatic cycles and of fluctuating desiccation.[1] He deduced that there was a period of intense aridity between the third and sixth centuries AD, during which time the southern silk route was abandoned in favour of the longer northern route; that between the ninth and sixteenth centuries there was a relative abundance of moisture, when rivers and population were much larger than today; and that from the sixteenth century onwards the climate has become progressively drier. He found evidence in support of this in the fluctuations in level of Lop Nor and of desiccation in the dissected condition of the lower slopes of the Kunlun, where the loess, once grass-covered, is now dead. The regions both north and south of Lop Nor, particularly in the Kuruk Tagh, formerly the grazing grounds of nomadic tribes, are now unable to support them through lack of water and grass. From Khotan eastward there are 13 notable rivers, some of which flowed as much as 25 miles farther into the desert than they do today, where there are ruins of former prosperous communities which would today need far better systems of irrigation than existed in their heyday. The condition of dead forests in the area points to their destruction being the result of decreasing rainfall rather than the cause. Owen Lattimore comments on this evidence that, while undoubtedly climatic changes have been the cause of great displacements of population in these regions, the abandonment of oases and irrigation systems may find their explanation in human activities such as inadequate engineering knowledge at the time to deal with increasing silting or increasing salinity or a community being reduced below survival point by war and raids.[2]

[1] See E. Huntington, *The Pulse of Asia* (Boston 1910), for a full discussion of these theories.

[2] O. Lattimore, 'An Inner Asian Approach to the Historical Geography of China', *Geographical Journal*, Vol. 110, Nos. 4–6.

Already enough mention has been made of inland drainage, of deserts and oases for it to be clear that this is a region of very arid climate. Rainfall is both very sparse and very variable from year to year. Kashgar has the heaviest rainfall with an average of 4·0 in. per annum, 2·4 in. of which falls in April, May and June. Everywhere within the basin experiences a big range of temperature (maximum shade temperature 115°F. in summer and a minimum of −20°F. in winter). The diurnal range is also very great owing to rapid radiation when skies are cloudless. As with other desert climates, dust haze often hangs over the landscape and dust storms, accompanied by fierce winds, are frequent.

Kashgar (4,003 ft.)	Jan.	Feb.	Mar.	Apl.	May	Jun.	Jly.	Aug.	Sep.	Oct.	Nov.	Dec.	Total
Temperature (°F.)	22	32	47	64	67	76	82	78	67	54	39	28	
Rainfall (inches)	0·2	0·2	0·2	0·6	0·8	1·0	0·2	0·2	0·0	0·6	0·0	0·0	4·0

However life in the Tarim Basin does not depend on the rainfall received, but on the irrigation waters descending from the vast reservoirs of ice and snow in the mountains to the north and south. In fact a 'heavy' rainfall in the spring months is something in the nature of a misfortune, since the clouds obscure the sun and retard the melting of the snowfields for which the rainfall is no adequate substitute. Although in 1931 Khotan received a heavier rainfall than usual, it suffered from a water shortage. Eighty per cent of Kashgar's water comes from snow and twenty per cent from springs. Its rainfall is regarded as negligible.[1]

Throughout the centuries life in the Tarim Basin has been based on arable farming and confined to the oases. Owing to the aridity and consequent paucity of grass, there has been very little nomadic pastoralism. About 85 per cent of the fields are irrigated; the distribution of water was formerly controlled by the local elders as it was in China, but it is now in the hands of the local authorities, usually communes. The main food crops raised are, in order of importance, wheat, corn, rice, kaoliang and barley. The main commercial crop is cotton. Usually local varieties are grown, because long staple USA cotton requires more water. Until recently only primitive agricultural methods and implements were used. The frost-free growing period averages 220 days.

This is one of the main regions of China chosen by the People's Government for major agricultural expansion and for the absorption of many millions of its fast-growing population. In 1932 Schomberg estimated that one third to one half of Sinkiang's water ran to waste, and that if this were

[1] Chang Chih-yi, 'Land Utilization and Settlement Possibilities in Sinkiang', *Geographical Review*, Vol. 39, 1949.

utilized, some 20 million acres could be brought into cultivation.[1] When civil war ended in 1949 the forces of the Sinkiang Military Command were formed into a Production and Construction Corps for the purpose of initiating land reclamation, which consisted mainly in building roads, reservoirs and irrigation canals, sluice gates, etc. This Corps, working in conjunction with the local peasantry, continues to function and has carried out a great deal of work of a capital nature, including the development of underground irrigation channels (*karez*). Claims are made that this has resulted in over 3 million acres of new land being brought under the plough, thus more than doubling the cultivated area in 1949. The main area of development has been along the northern side of the basin. New crops, notably sugar beet, are being grown. Grain crops have been doubled, whilst cotton, owing to its tolerance of saline soils, has been increased eightfold. The dangers facing this rapid increase are those of former centuries, i.e. shifting sands and the development of alkaline soils. It remains to be seen whether modern techniques and know-how can adequately meet these dangers. No railway has yet been built to serve the Tarim Basin but an extensive road system is in the process of being developed.

The population of the whole of Sinkiang according to the 1953 census was 4,873,608, of which about 75 per cent live in the Tarim Basin. About 80 per cent are Uighurs, who are Moslems of Turkic stock, and some 9 per cent are Kazaks, who are also Moslem. Only 5 per cent are Chinese. The remainder is made up of various nomadic tribes of Torgut, Noigut, Kirghiz, Kalmuks and Tungus, who live on the southern slopes of the T'ien Shan. The Uighurs form a homogeneous group of independent people. For this reason they have been given the somewhat more independent status of an autonomous region and have been treated circumspectly by the central government, which has not pressed socialization quite so hard or so rapidly as in other parts of China. Nevertheless, it is reported that from the two and a half million acres of land reclaimed by the Production Corps since 1958, 102 large new mechanized farms have been established.

7b. *The Turfan Depression*

To the north-east of the Tarim Basin and lying between the Kuruk Tagh or Dry Mountains and the Bogdo Ola is the remarkable Turfan Depression. This is a fault trough, which descends to 505 ft. below sea-

[1] R. C. F. Schomberg, 'The Habitability of Chinese Turkestan', *Geographical Journal*, Vol. 80, 1932.

level. It consists of two main valleys of different height divided by a faulted range, the Fire Mountains, which, in spite of their name, are not volcanic. The wide valley to the north of the Fire Mountains is above sea-level and some 1,500 ft. higher than the valley to the south. It is covered with a thick deposit of piedmont gravel, ground water from which drains by way of canyons through the Fire Mountains into the lacustrine plain below. This valley is bounded on the north by the fault face of the Fire Mountains, giving a range of 2,000 ft. in elevation on its southern face. A belt of piedmont gravel lines the southern border at the foot of the Dry Mountains, alongside which is also the salt playa or swamp, the remnants of the former lake which filled the depression. Today this swamp is dry in summer and wet only in winter when it freezes and when evaporation is not as great.[1]

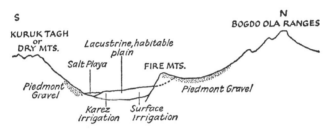

FIG. 77. Section through the Turfan depression

This lower plain is the habitable part of the Turfan Basin. It has no river of any size but a large number of streams from the Fire Mountains, which provide surface irrigation on the northern side of the plain. As the plain dips to the south, surface irrigation gives place to the *karez*, the Persian and north-west Indian method of long underground tunnels, which tap the water table and lead the water to the fields.

Turfan has a climate of great extremes of temperature. Summers are intensely hot, shade temperatures rising to 125–30°F. at midday. Every house has its dug-out to which its inhabitants can retire for siesta until work is again possible. Winter temperatures fall below zero. Rain in the depression is practically unknown.[2]

All crops are grown by irrigation, the chief being fruit, cotton, grain and silk, in that order of importance. Turfan's fruit is famous, especially its grapes. From its vineyards come seedless grapes which make high quality

---

[1] E. Huntington, op. cit.

[2] M. Cable and F. French, *The Gobi Desert* (London 1946).

sultanas: currants, raisins, prunes are all sun dried. Other fruits are apricots, peaches, melons and nuts. As in China Proper, the farmer never fails to grow peas and vetches around his grain plots and vineyards, thereby deriving not only the crop itself but adding nitrogen to the soil.

As compared with the rest of China, the Turfan Basin has hardly been touched by the changing order. Nevertheless, farming by 1956 had been organized into co-operatives and presumably later into communes. The people are of Uighur stock and are Moslems. Many ruined towns and villages dot the region.

7c. *Dzungaria*

A vast basin, some 2,000 ft. lower than the Tarim, lies to the north. It is triangular in shape and is contained within the walls of the T'ien Shan on the south, the Altai Mountains on the north-east and the Ala Tau on the north-west. A pass over the col between the Bogdo Ola and the T'ien Shan at a height of about 12,000 ft. leads from the Turfan Depression to Urumchi and the Dzungarian plain. The Dzungarian Gate, a rift valley in the Ala Tau, provides an outlet from the plain to Lake Balkhash and Kazakstan. The upper waters of the Irtysh River give access to Zaisan Nor and Semipalatinsk in the USSR, while the wide valleys between the Bogdo Ola and the Altai provide a way, albeit a dreary one, in the south-east to the Mongolian Gobi.

The region has much the same pattern as the Tarim Basin but at a lower level, with the difference that it is not so arid. The central part is desert but not of the same expanse of bitterness as the Taklamakan. The two main rivers are the Manaas and Urungu. These, in company with innumerable streams, descend to the plains to be lost in the reedy marshes and in the sands. The high mountain ranges receive as much as 20–30 in. of rain, and those ranges of the north, notably the Altai, are forest-clad. Their lower slopes are wooded with willows, poplars and alders, above which come white birch, a valuable building material and one which has many uses among the nomads. Its bark contains oil which is used by the herdsmen as a lining to their pails for sour milk. Above 6,000 ft. the Siberian larch dominates. These trees grow to over 100 ft. in height and form 70 per cent of the forest growth. These forests are rich in fur-bearing animal life—fox, wolf, sable, ermine, bears and wolverines.

In contrast to the Tarim with its desert lands between oases, steppe and steppe-desert are the natural vegetation of the wide valleys which border the central desert-land. In consequence pastoral nomadism has been the dominant way of life in Dzungaria. Population is sparse since the pasture will not carry many sheep to the square mile. According to Hann a rainfall of 20 in. a year in New South Wales makes it possible to keep 600 sheep on a square mile of land. With a rainfall of 13 in. only 100 can be kept and with only 10 in. only 10 sheep per square mile. Here in Dzungaria the annual rainfall is under 10 in. with the result that the pastoralist must be constantly on the move in search of fresh grass. This fact emphasizes a further contrast between the Tarim and Dzungaria. The routes in the former are fixed, going from oasis to oasis and their purpose is primarily trade. In the latter they are rather what Lattimore calls 'directions of march', broad and indeterminate ways along which the herdsman leads his flocks.[1] Trade is entirely subordinate to pasture.

Until the People's Government came into power, very little arable farming was practised. Since 1949 strenuous efforts have been made towards its development. Here, as in the Tarim Basin, the main initiators and enthusiasts have been the members of the Production and Construction Corps. Operating mainly along the foothills of the northern slopes of the T'ien Shan from which the Manaas and several other smaller rivers flow, big water conservation works have been carried out and reservoirs and irrigation channels built, thus saving much of the water which formerly ran to waste. By 1956 the Corps had reclaimed over 125,000 acres, which have been organized into ten large mechanized farms. Since that date work has gone ahead at an even greater rate. About half the reclaimed land is under cotton; the rest is under rice, wheat, corn, soy-bean, fruit and sugar beet.

However, economic interest has centred not so much in agricultural advances, great though they are, as in the discovery in 1955 of oil in a large field at Karamai, 200 miles north-west of Urumchi. The first attested field was 23 sq. miles in area with a reserve of 100 million tons. Since 1955 intensive surveys have been carried out and six new oilfields discovered. A pipeline has been laid to Yumen and Lanchow where there are now large refineries. The Yumen Oil Administration, which manages the field, had a production target in 1959 of 100,000 tons and the processing of 30,000 tons.

The discovery of oil in distant Karamai has quickened the building of the trans-Mongolian railway from Lanchow to the Russian border. The line follows precisely that of the old Imperial Highway as far as Hami; then it goes to Urumchi and to the Dzungarian Gate. The building of this line has been fraught with difficulties, partly owing to the very unfavourable

[1] O. Lattimore, 'Caravan Routes in Inner Asia', *Geographical Journal*, Vol. 72, 1928.

terrain of mountain and shifting sand, and also in part due to the great haste attending its construction.

Urumchi, or to give it its Chinese name, Tihwa, stands on a desert plateau of 9,000 ft. and is the capital of Sinkiang Uighur Autonomous Region. It has grown in a matter of ten years from an old market centre of 60,000 inhabitants to a busy modern commercial and industrial city of 150,000. In this time metallurgical, chemical and machine works, power and cement plant have been built, coal mines opened and oil refineries constructed. Much energy has been put into education. Many schools and colleges have been built but there is a great shortage of teachers. Urumchi is essentially a Uighur town, Uighur being the official language. There is a Kazak sub-district of the city. The population is almost entirely Moslem.

The Ili Valley

The small upper Ili Valley lies within Sinkiang's borders. Its funnel mouth faces due west and in consequence is the recipient of a more generous amount of rain than any other part of the province. The resultant prosperity of this good fortune led Little to describe the Ili Valley as 'the gem of Chinese Turkestan'. It is thickly settled by Taranchis, i.e. agriculturalists. It also raises large numbers of sheep, cattle and horses.

8. INNER MONGOLIA

The present political boundaries of Inner Mongolia lie between the southern border of the People's Republic of Mongolia, i.e. Outer Mongolia on the north and China Proper on the south, reaching well into the Hwang-ho Basin to include the Ordos. Recently it has swallowed up the former provinces of Ningsia, Chahar and Jehol. Its capital is Huhehot, formerly known as Kweihwa. Marco Polo refers to it as Tenduc.

Geographically its southern boundaries within China are formed by the series of high ridges of the Ala Shan, Hara (Kara) Narin Ula and the Ta Hingan Shan. This series of mountain ranges form the high southern rim of the Gobi depression. They have an average height of 4,500 to 6,000 ft. and in many places they rise to 7,000 to 8,000 ft. The land falls away to 2,000 to 3,000 ft. towards the centre of the Gobi, which is the Mongolian name for a broad desert plain. This region of inland drainage and salt marshes is by no means uniform in character; stretches of coarse sand give place to areas of piedmont gravel and smooth rock and stone.

The deeper the Gobi is penetrated northward and westward, the drier it becomes. Rainfall is more precarious, variable and sparse. When rain does fall, too often it is in the form of a violent cloudburst. It is a region of great extremes of temperature, both seasonal and diurnal. Midday shade temperatures in summer reach 115°F. and more. In winter 50–60° of frost are experienced. These changes in temperature give rise to terrifying sandstorms and violent windstorms, which, in winter, are bitter and penetrating.

The mountains on the southern rim have some cover of woodland and scrub but vegetation for the most part is very scant. Some of the Gobi ranges are entirely barren. As the land dips away northward, in so far as it has any cover this is steppe grass, which sometimes stands 7 to 8 ft. high, looking like kaoliang except that it carries no heavy head of grain and is slenderer in its stalk. Like bamboo in southern China, this grass is put to a multitude of uses. The young, tender shoots are fodder for the livestock. Fully grown, it is used to make kitchen utensils, brushes, grain bins, mats for beds, curtains screens, toys, chassis for carts and fences for fixing sand dunes.[1]

The region is one of very sparse population. Over the whole area there is an average of less than one person per square mile. Three categories of people make up the inhabitants. There are the oasis dwellers, a settled people, the great majority of whom never move outside the limits of their small world from birth to death. Here they till their fields and gardens, the produce of which is often unique to the particular oasis, and pursue their individual handicrafts. Their products are distributed by a second class of people, the caravaneers, who move along the desert 'highways' from oasis to oasis and from water-hole to water-hole. These latter are usually about 30 miles apart, a distance which can be accomplished in one day. The region is still virtually untouched by modern transport and nearly all movement is done by camel, mule and mule cart, horse or on foot. The pastoralist forms the third category of inhabitants. For the most part the Mongolian pastoralist is only semi-nomadic. He practises transhumance, that is to say, he makes seasonal movements with his flocks and herds but prefers to restrict his movements as far as possible. However, he is a *yurt* dweller and therefore can and does move as necessity dictates. The average *yurt* is circular and is from 12 to 15 ft. in diameter, having wooden trellis walls, usually of willow, about 4 ft. high. The whole is covered by felt, the

[1] M. Cable and F. French, op. cit., p. 99.

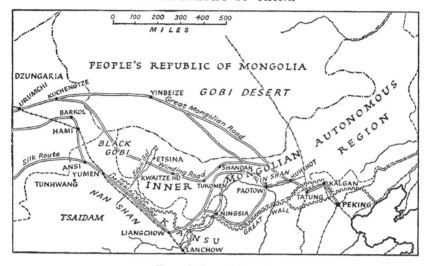

FIG. 78. Inner Mongolia

roof being upheld by long sticks of willow joined in the centre. It provides snug protection in the bitter winter. It is women's work both to pitch and strike camp, the packing being a comparatively simple job. The men are responsible for herding and hunting. The daily food of the pastoralist is derived directly from his flocks of sheep and goats and herds of camels, horses and cattle. Milk, cheese and rancid butter together with mutton form his staple diet. His drink is tea, brought in brick form from Central China and usually concocted with mutton broth.

Since 1949 a widespread and determined attempt has been made to bring these semi-nomadic people into line with the communist pattern of the rest of the country. They were first organized into more than 2,000 herdsmen co-operatives, which were later merged into 152 communes. It is clear, however, from later reports that considerable latitude and flexibility were allowed. Commune members are allowed to keep for their private use the necessary number of horses for personal transport and milch cows and sheep for food. Nor does the commune attempt to under-take the full supply of food needed by its members.

'A clear distinction must be made between means of production and means of subsistence and different policies must be laid down for dealing with them. Members should be told clearly that they still retain, and will always continue to retain, private ownership of their means of subsistence (including houses, *yurts*, furniture, clothing, rugs, saddles and trappings for horses, etc.) as well as their deposits in the bank or in credit co-operatives, their cash and their personal ornaments made of gold and silver.'[1]

Some attempt is being made to develop agriculture in the land lying east and west of Huhehot, on the inside lip of the Mongolian Basin and north of the Great Wall. This is part of that 'belt of contention' between the agriculturalist and pastoralist.

The main east-west line of communication running through Inner Mongolia from Huhehot to Turfan and Urumchi is the Winding Road, described by Owen Lattimore.[2] It is of less importance today than the Great Mongolian Road to the north and the Imperial Highway to the south, along which now runs the railway line from Lanchow to Sinkiang. Its former importance is attested by the ruined remains of Etsina on Edsin Gol, a walled city inhabited in Marco Polo's day but long since deserted, and by the two big lamaseries of Shandan and Tukomen. Monasteries were usually sited where nomads gathered for festivals. The ruins of Estina may possibly be evidence of progressive desiccation in this region and of an Edsin Gol of considerably greater volume then than now. This river provides the only practicable north-south route across western Inner Mongolia and is the one which Genghis Khan followed in his invasion of China in AD 1227. West of Edsin Gol the Winding Road crosses the Black Gobi, the most forbidding of all the Mongolian deserts. It is a plateau of hard, sandy clay; it is practically rainless and has no wells. The Winding Road experienced a revival of popularity in the 1920's and 1930's when use of the northern Great Mongolian Road was denied by an independent and hostile Outer Mongolia and the Imperial Highway was cut by civil war and banditry.

9. THE TIBETAN PLATEAU

The Tibetan Plateau is today divided politically into three provinces: Tibet itself, Chamdo Area and Chinghai. These three roughly correspond to three natural regions within the plateau.

[1] *Jen Ming Jeh Pao*, Peking, 25th January 1959.
[2] O. Lattimore, 'Caravan Routes in Inner Asia', *Geographical Journal*, Vol. 72, 1928.

9a. Tibet

Tibet is a vast area of upland, having a general level of 16,000 ft. or above, from which mountain ranges rise a further 10,000 ft. or more. Its borders are marked by great mountain systems which stem from the Pamir Knot in the west. In the north the Kunlun and Astin (Altyn) Tagh divide it from Sinkiang below and on the south is the great Himalayan system, cutting it off from the Indo-Gangetic plain. In between these are other ranges, which run roughly parallel to the bordering ranges. The Tangla Mountains are a continuation eastward of the Karakoram, and south of the Tangla Ranges are the Trans-Himalayan Ranges.

Tibet can be sub-divided into two main areas. The north and north-west, known as Chang Tang, is the highest part of the plateau and is a land of wide, barren, desolate valleys along which sweep the bitter winds so feelingly described by Peter Fleming in his *News from Tartary*. It is a region of inland drainage. Its streams lose themselves in innumerable salt lakes, the largest of which is Tengri Nor, and in salty swamps, which lie in salt-encrusted plains. The natural vegetation of the north and north-west is alpine desert, for it is treeless. In some parts there is a growth of scanty grass, which provides sustenance for the wild mountain sheep, asses and yaks. Human population is very sparse indeed. Not surprisingly, there are no adequate climatic records of this remote and inhospitable land. Kendrew,[1] however, makes estimates which give some idea of the rigours to be endured. At 16,000 ft. the mean temperatures are −10°F. in January and 30°F. in July. Most of the region has less than two months with a mean temperature above freezing. In such a climate arable farming is impossible.

The Chang Tang has a general slope south and south-east down to the Trans-Himalayan ranges, south of which is a region known as Po. This is a trough lying between the Trans-Himalayan ranges and the Himalayas themselves through which the upper courses of the Indus and Brahmaputra (Tsangpo) flow. Here are deep valleys, some of which are less than 5,000 ft. above sea-level. Temperatures are a good deal warmer than on the plateau proper, but even so are cooler than south-east China. There has always been some cultivation here but it is only since the People's Government forced its way in and ousted the régime of the Dalai Lama that there has been any large scale development of arable farming. Highland barley is the staple food. Po has a wealth of forest growth denied to the rest of Tibet. There has recently been some development of light industry (clothing, footwear, woollen textiles, leather, timber) in the Lhasa area, using hydro-electric

[1] W. G. Kendrew's chapter on Climate in Dudley Buxton's *China, the Land and the People* (London 1929).

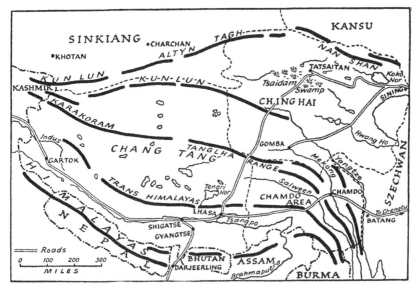

FIG. 79. Tibet

power. One of the reasons for the Chinese Government's anxiety to bring Tibet, especially Po, firmly under its jurisdiction may be the potential mineral wealth which lies hidden in this part of the plateau.

Po is the only part of Tibet that has any density of population. Three great centres of Lamaism in Lhasa, Shigatse and Gyangtse are located in this south-east corner on which the old caravan routes and now the two new motor roads converge.

9b. *Chamdo Area*

The former province of Sikang has been swallowed up by the extension of Szechwan westward to the banks of the upper Yangtze or Kinsha and by the creation of a new administrative area, known as Chamdo Area. Chamdo embraces the headwaters and the upper basins of the three great rivers: the Yangtze, Mekong and Salween. It is the region where the great folds bend south-east and then south, between which the rivers flow in the same deep and forested chasms already described when dealing with the South-west Plateau. It is still largely unexplored and is inhabited by only a few aboriginal tribes.

Travellers and surveyors returning from this area tell of its mineral wealth but its inaccessibility makes exploitation in the near future unlikely. However, it is being opened up to some extent. The old routes from Lan-

299

chow via Sining to Lhasa, and from Chengtu via Chamdo to Lhasa have now been sufficiently built and engineered to permit motor traffic.

9c. Chinghai (Tsinghai)

The province of Chinghai contains the basin known as the Tsaidam, which lies at a much lower level (9,000–6,000 ft.) than the rest of the plateau. It is bounded on the north-west by the Astin (Altyn) Tagh, on the north-east by the Nan Shan, a branch of which separates Koko Nor (Chinghai) from the Tsaidam itself. Ranges of the Kunlun mark its southern borders. Both the Tsaidam and the Koko Nor areas are regions of inland drainage.

In the centre of the western end of the Tsaidam there is some desert with a periphery of swamps and salt lakes. On the rising rolling land which surrounds them, reasonable steppe pasture is found. On the slopes of the Astin Tagh and Nan Shan the grass is really good over large areas and the higher slopes carry some forest.

Through the centuries this remote region supported a sparse population of Tibetan nomads who were an ever-present menace to the caravans following the Imperial Highway north of the Nan Shan, and sometimes a threat to imperial power. This area has suddenly sprung into a new prominence. Geological surveys during the last 12 years have revealed the presence of both oil and coal. The main oilfield lies in the extreme west, having Mangyai at its centre. A second field is sited at the heart of the region at Tatsaitan. Close by Tatsaitan, at Yuka, whose Mongolian name is Naka, meaning 'thick forest', is a coalfield with seams 125 ft. thick, lying near the surface. Big iron ore deposits have recently been discovered in the Nan Shan area. In the east, near the settlement of Chaka, is a crystallized salt lake with deposits of 98 per cent sodium chlorate 50 ft. thick over an area of 40 sq. miles. This will probably be the centre of a chemical industry in the near future.

Chinghai is noted for its horse breeding. Legend has it that the horses were 'dragon-bred', piebald in colour and of phenomenal staying power, capable of doing 1,000 li (more than 300 miles) a day! We do know of their great sturdiness from accounts of the amazing surprise rides accomplished by Jenghis Khan's men and there are plenty of records today of horses trotting more than 30 miles in a day. It is reported that there are now some 350,000 horses in the area.

Sining wool is famous. It is very long staple, coarse, elastic and strong, much used in rug and carpet making. Two medicinal products in big

FIG. 80. The Tsaidam Basin

demand in China are Chinghai rhubarb and *tungts'aochung* (winter grass insect).

The whole province is sub-divided for administrative and governmental purposes into six autonomous *chou*. The capital is Sining but Tatsaitan appears likely to become the main settlement and administrative centre in the rapidly developing western part of the province. In order to open up the region a railway has been run from Lanchow to Sining and may already have been completed as far as Tatsaitan. A motor road runs from Sining, via Chaka, Tatsaitan and Yuka, over the Tang Ching Pass between the Astin Tagh and Nan Shan to join the main route to Sinkiang through the Kansu 'pan handle' at Ansi.

Population in Chinghai had already risen to 1,676,500 in 1953 and, although statistics are not yet available, it seems reasonable to suppose that this figure was more than doubled within a decade. Although the climate here is still harsh, summers are sufficiently long to permit some cultivation, given careful water conservation. The area has promising prospects for the development of pastoral farming on a large scale.

10. TAIWAN

Perched, as it were, on the brink of a submarine chasm stands the island of Taiwan. It forms a bastion in one of the great festoons of islands which fringe the western Pacific and which are part of the Alpine system of folds ringing the globe. Like Japan, it stands on the edge of the continental shelf, the sea descending to abyssal depths within a few miles of the eastern coast, while the depth of the Formosa Channel separating the island from the mainland is a mere 150–250 ft. As a result of intense folding and faulting in both pre-Tertiary and Eocene times, the island forms a tilted block, dipping towards the mainland and having an axis of folding from NNE to SSW, extending the whole length of the land.

The island, for the most part, is very mountainous and the effect of the trend lines is to divide it into a series of five longitudinal physiographic regions.[1] On the east coast, between Hualien and Taitung, are the Taitung Mountains, rugged and so steep as to make communication along that stretch of the coast virtually impossible. These mountains are composed mainly of Miocene volcanics with some sandstones and shales. Their western edge is a precipitous scarp, which forms the eastern boundary of a rift valley, whose opposite scarp is the western edge of the Central Mountain Ranges. This valley, known as the Taitung Rift, about 100 miles in length and on average 4 miles in width, is filled with recent alluvial deposits. The third region, the Central Mountain Ranges, extends throughout the central length of the island. They consist of three main ranges: the Changyang Mountains, Yushan and Alishan, named from east to west. They are composed mainly of schists, quartz and gneiss and form a very imposing landscape, having 27 notable peaks, many of which rise to over 10,000 ft. Much of the Alishan is faulted, giving rise to depressions, in one of which lies the beautiful Sun Moon Lake. To the west of these Central Ranges lies a wide belt of hilly country whose terraces bear evidence of at least two uplifts and the rejuvenation of the river system. In the centre these hills embrace a small alluvial basin in the heart of which Taichung stands. Again west of the hills an alluvial plain runs the whole length of the island. In the centre it is 30 miles wide. Across this plain flow many rivers, the four most notable of which are the Tachia, Talu, Hsilo and Tsengwen. While

[1] Vei Chou Juan, 'Physiography and Geology of Taiwan'. This and other pamphlets of the Chinese Culture Publishing Foundation, together with Chang Jen-hu's 'Agricultural Geography of Taiwan' have been widely drawn on for the material of this chapter.

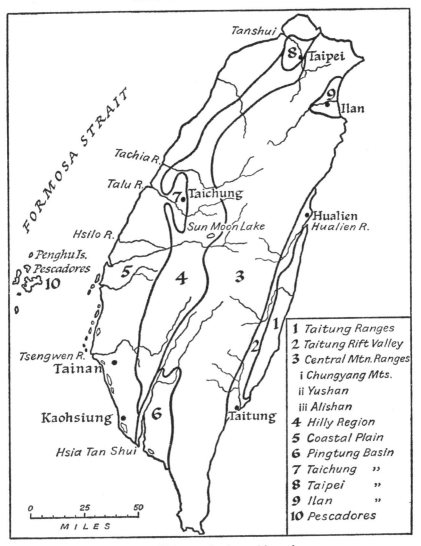

FIG. 81. Taiwan: physiographic regions

they are torrential in their upper courses, their gradients across the plain are such that they become meandering and are liable to change course, building deltas and lagoons at their mouths. There is one large river in the south, the Hsia Tan Shui, which follows a longitudinal course between the north-south folds and which has built up an alluvial plain of sands, clays and gravels in the south. This is known as the Pingtung Plain. Two

303

smaller alluvial basins in the north, formed by the Tanshui and Chu Shui, have Taipei and Ilan in their centres respectively.

A group of islands, known as the Pescadores or Penghu Islands, lies 25 miles off the west coast in the Formosa Strait. These islands are built of a basaltic flow, 'a fragmented and dissected mesa . . . composed of three flat-lying basalt flows'. There are 64 islands in all, many of which are fringed with coral reefs.

Taiwan has an area of 13,800 sq. miles or about 36,000 sq. km., about the size of Holland or a little smaller than Switzerland. Its coastline is exceptionally smooth and unindented, being only 700 miles in length. It has no good natural harbours. The southern half of the west coast is shallow and full of lagoons; the east coast is steep, smooth and without inlets.

The Tropic of Cancer passes through the middle of Taiwan, which stretches from 20° 53′ N. to 25° 18′ N. It thus enjoys a climate very similar to that of Kwangtung with modifications due to its being an island 100 miles off the mainland. With the exception of the north and north-east, the rest of the island has the normal monsoon pattern of summer rain and winter drought. The outflowing cold air from the Siberian high in winter passes over the East China Sea before it reaches Taiwan. In that passage it becomes warmer and picks up a good deal of moisture. It strikes the island from the north-east and brings to that part quite a heavy winter rain, while the rest of the island is in a rain shadow. Chilung, in fact, has a winter maximum of rain and a fairly even distribution throughout the year; Ilan on the north-east coast has a maximum in October. Between June and September the winds, which are generally lighter than those of winter, blow from the south-west, bringing a marked summer maximum to the south and west coastal plains. Summer, too, is the typhoon season and Taiwan, to its sorrow, lies right in the most used paths of these unwelcome visitors. On an average three typhoons pass directly over some part of the island each year and many others pass nearby. They bring heavy rainfall in addition to widespread destruction to crops, buildings and shipping. Temperatures on the plains are sub-tropical or tropical. As in Kwangtung, summers are long, hot, humid and enervating. Mid-day summer temperatures are usually over 90°F. and relative humidity is often between 85 and 90 per cent.

The natural vegetation of the whole island would be tropical and sub-tropical forest, but as with the lowlands and plains of China, most of the western coastal plain has been cleared of its tropical woodlands to make room for cultivation. Mangrove swamps occupy some of the tidal deltaic areas. Most of the more mountainous region, however, still retains its magnificent cover of forest. On the lower slopes, up to about 5,000 ft.

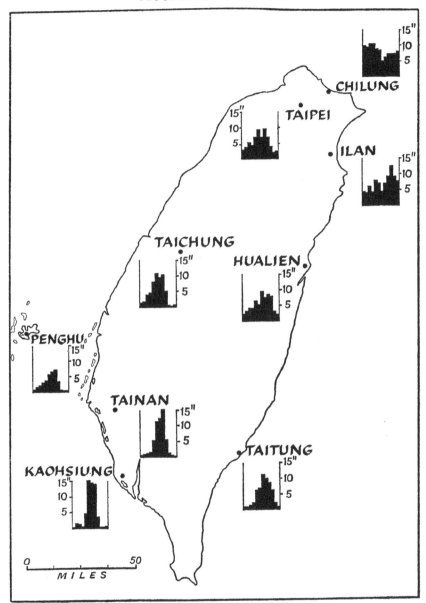

FIG. 82. Taiwan: rainfall map with graphs

broadleafed evergreens flourish, including the camphor tree and oak. As elsewhere on the mainland, the camphor tree, because of its commercial value, has suffered heavy cutting, which, although curbed by the Japanese during their occupation, has resulted in bad soil erosion. Above 5,000 ft. the broadleafed evergreens give place to conifers, pine, larch, cedar and still higher fir. It is estimated that more than half of the island is still forested. Given careful forestry and conservation, this is a potential source of considerable wealth.

The resulting soils are similar to those of south China, i.e. mature, leached, acid, in need of constant fertilization and liming. On the plains there is the same tendency to laterization and on the hills to podzolization.

Although the existence of Taiwan seems to have been known to the Chinese in the seventh century, this knowledge appears to have faded from consciousness much as did the realization of the existence of North America disappear from European ken for centuries and had to be rediscovered. In 1430, not so many years before the great discoveries in the West, Admiral Cheng Ho, returning from an expedition to Thailand, was driven off course and landed on Taiwan. He returned home and gave glowing accounts of this land he had discovered. This marked the beginning of Chinese colonization of the island, which went on intermittently until interrupted for a while by Western activities. Early in the seventeenth century the Portuguese landed and made their first settlements there. It is from them that the island got its name Formosa (Ilha Formosa, meaning 'the Island Beautiful'). The Chinese name, Taiwan, is composed of the two Chinese words *t'ai*, meaning 'look-out tower' or 'platform' and *wan* meaning 'bay'. The Portuguese were followed by the Dutch and the Spaniards, who strove to drive the Portuguese and each other out. Victory fell to the Dutch in 1642, but their triumph was short-lived.

Just at this time the Manchus were invading China and ousting the Mings. Among the many leaders and generals attempting to stem this onslaught was Cheng Cheng-kung, better known to the west as Koxinga. After fighting a losing battle against the Manchus, he shipped his army to Taiwan and there summarily defeated and drove out the Dutch (1661). After his victory many Chinese, estimated at over 2 million, who were unable to stomach Manchu domination, fled to the island. Koxinga ruled Taiwan as an independent country till his death, but in 1683 it fell to the conquering Manchus and was incorporated into China as a *fu* or district of the province of Fukien. Thereafter it continued its intermittent and undirected colonization and development, which was attended by endemic warfare with the Malayan aborigines, who were forced back into the mountains where their remnant of not more than 150,000 are still to be found.

Western and Japanese interest in Taiwan again revived in the second half of the nineteenth century as realization of its worth as a market and a source of raw material began to dawn. In order to forestall any annexation by the Great Powers, the island was again incorporated into China and given the status of a province in its own right in 1887. This precaution was, however, fruitless, for in 1895 Japan claimed the island as the fruit of victory in the Sino-Japanese War (1894–5) and for the next fifty years it was ruled by that country as part of its empire.

Those fifty years saw great development in its economy. Road and rail communications were built the length of the western coast. Water conservancy and irrigation schemes and hydro-electric works were carried out with a view, as later in Manchuria, to meeting the ever growing needs of the homeland in Japan. It was during this period that the great hydro-electric generating station at Sun Moon Lake was built. Crop land was increased and the production of all foodstuffs, particularly rice and sugar, stepped up. Undoubtedly this period of Japanese rule was one of great economic development.

With the defeat of Japan in 1945, the island fell under the rule of the Kuomintang. As many of the Taiwanese did not take kindly to their new rulers, three or four years of harsh and unsettled government ensued, culminating in 1949 in the establishment of Taiwan as the seat of the Nationalist Government under Chiang Kia-shek when ousted from the mainland. In that year and ever since there has been an influx of Chinese into Taiwan. The 1951 census placed the total population at 7,791,000 of whom more than 2 million were immigrants of the previous five years. Except for the handful of aboriginals in the mountains and some 180,000 peasants in the rift valley, practically the entire population is distributed densely along the western coastal plain and in the adjoining hilly district. It is estimated that there are 2,527 persons per sq. mile of cultivated land and more than 600 per sq. mile for the whole country, which is a greater density than that of the mainland. This population is essentially rural. More than 90 per cent is engaged in agriculture, either directly or indirectly. Taipei, the capital, is the only town with over 500,000 inhabitants and there are only three others—Taichung, Tainan and Kaohsiung—with a population of more than 200,000.

When the island was taken over by the Nationalist Government in 1949, the need for land reform and some relief for the peasant was no less urgent here than on the mainland. In view of the diametrically opposed ideologies of the Kuomintang and the Communists it is not surprising to find that the approach to the problem has been very different. In Taiwan, although the landlords have been compulsorily relieved of their lands, they have been compensated quite liberally by being given holdings in the stock of one of five great government owned corporations: Taiwan Cement Corporation, Taiwan Paper and Pulp Corporation, Taiwan Engineering and Mining Corporation, Taiwan Agricultural and Forestry Corporation and the Taiwan Fertilizer Company. The fact that this compensation has been possible is due in no small degree to the financial support which the island has received from the USA. The peasants who received the redistributed land have been organized into co-operative societies for collective buying, selling and credit.

Of all the crops grown, rice is by far the most important, both in value and in acreage. More than four times as much land is devoted to it as either sugar cane or sweet potatoes, which are its nearest rivals. Moreover about 60 per cent of the rice grown is double cropped, which is tantamount to a further increase in acreage. Some 1·42 million metric tons were produced in 1950. In value it represents nearly 44 per cent of the total value of all agricultural produce. It is grown throughout the whole length of the western coastal plain, relying almost entirely on the vast system of irrigation for which the Japanese are mainly responsible. A very small percentage of upland rice is grown. Wheat is the second cereal grown, its production being confined to the southern half of the plain. Like so much of the rest of China, the sweet potato forms an important item of diet for the poorer sections of the community. About 2·35 million metric tons were grown in 1950, and although its cash value was only one eighth that of rice, this fact should not be allowed to obscure its real value in the country's economy.

Sugar cane, introduced by the Dutch in 1642 and greatly encouraged and developed by the Japanese, is the most important cash crop. Its growth is concentrated in the southern half of the coastal plain where the climate is hotter. It formed one fifth of the total value of agricultural produce in 1950. Second to sugar as a cash crop is tea, the growth of which is confined almost entirely to the slopes of the northern hills, where the more even distribution of rainfall is conducive to the better growth of the leaf. Recently there has been an increase of plantations in the hills of the Sun Moon Lake region.

Fruit grows prolifically on the hills and plains of Taiwan. Banana trees are ubiquitous and are the chief fruit grown. The big plantations are located in the Taichung Basin and on the plains to the south of it. A large part of the crop is exported to Japan and Hong Kong. Pineapple is another fruit which is grown extensively, the main concentration being again in the Taichung region. Other fruits are mango, citrus (pomelo, tangerine, grapefruit, lemon), papaya, pomegranate, persimmon and peach.

In addition to the usual Chinese fibres of hemp and ramie, some sisal and jute are grown on the coastal plain south of Taichung. Although the production of jute amounted to only 9,600 tons in 1950, it did, in fact, rank third in world production.

The coastal formation does not lend itself to the development of good natural harbours and this fact has its influence on the fishing industry, which is not so important a factor in the lives of the people of Taiwan as it is on those of Fukien just across the Formosa Strait. Nevertheless, more than 150,000 of the Taiwanese are engaged directly or indirectly in fishing. Recently there has been a big extension of mechanization of the deep-sea fishing fleet, and some development in freshwater pond fisheries.

Taiwan is not well endowed with mineral wealth. There are some young coal measures in north-western Taiwan, laid down in the Oligocene and Miocene, which are mainly lignite. There is very little coking coal and no appreciable reserve of iron ore on which an iron and steel industry could be based. There is, however, as we have seen, an abundance of hydro-electric power and on this a small aluminium industry at Kaohsiung has been built. Bauxite is imported from Indonesia. Production in 1952 amounted to 3,856 ingot tons.

There is a flourishing paper and pulp industry which draws its raw material largely from bagasse, the waste of the sugar cane industry after the sugar has been expressed. Some 10 tons of bagasse, the result of producing 7 tons of sugar, will in its turn produce 2½ tons of paper, which is used in making bags to serve the cement, sugar and fertilizer industries.

There is a very small production of oil in the north. There is little hope of discovering large reserves of oil in the island on account of the heavy folding and faulting of the strata, which has destroyed the anticlinal formations in which oil might be trapped.

The excellent limestone and silica sand, found in plenty, has given rise to a flourishing cement industry, which in 1952 produced 450,000 tons.

Practically all the industry which today exists in Taiwan owed its origin and development to the Japanese during their fifty years of occupation. After the war, ownership passed into the hands of the Chinese government. Small concerns were sold to private firms and individuals, but the large corporations, which have been named in connection with landlord compensation and land reform, have remained in government hands. Towards the end of the Second World War industrial plant suffered very heavily at the hands of Allied bombers. As a result, as in Manchuria, reconstructed industry is running largely on the newest equipment.

The geo-politics of Taiwan are at present difficult and involved. While neither the Communist Government of the mainland nor the Nationalist Government of Taiwan is in any doubt that the island is an integral part of China, the former regards it as the last remaining 'unliberated' part of the Fatherland still in the hands of the 'running dogs of the imperialists' and the latter regards it as the refuge from which the rightful Pretender will one day return to the mainland to re-establish a free China. Behind all this lies the USA, which regards Taiwan as an essential bastion in its western Pacific fringe of defences against the communist menace of China and the USSR.

11. HONG KONG AND THE NEW TERRITORIES

The bargain which Captain Elliot struck with the Chinese government at the end of the 1841 war for the cession of Hong Kong to the British in return for the Chusan Island, captured during hostilities, was regarded generally as a poor one. Palmerston, Foreign Secretary at the time, dismissed Elliot from his post with the comment that he 'had obtained the cession of Hong Kong, a barren island with hardly a house upon it . . . It seems obvious that Hong Kong will not be a mart of trade any more than Macao is.' Palmerston was right in his description of the island: it was bare, treeless and rocky with only about 2,500 inhabitants; but he was quite wrong in his prognostication. In spite of many vicissitudes, epidemics and fevers, storms and typhoons, commercial crises and human blunders and a good deal of opposition both at home and on the spot, the colony has grown with great rapidity over the last 120 years.[1]

The earliest settlement was at Victoria on the north-west of the island. This city has now spread the whole length of the north-west coast and numbers over one million inhabitants. In 1861 the peninsula of Kowloon on the mainland was ceded by China, thus adding a further 4 sq. miles to the 32 sq. miles of the island. This addition, together with Stonecutter Island, enabled the British to control adequately the magnificent natural harbour, which gives 17 sq. miles of deep and safe anchorage. Kowloon at that time consisted of a Chinese fort, which occupied the extreme south promontory and which had been a constant danger and annoyance to shipping in the harbour. There were also several small villages on the peninsula. Probably the population was less than 3,000. Today it is a thriving, densely peopled city of 1½ million.

[1] T. R. Tregear and L. Berry, *The Development of Hong Kong as told in Maps* (Hong Kong 1959).

In 1898 a further 360 sq. miles were added to this colony of only 36 or so sq. miles on the plea that more land was necessary for the 'proper defence and protection of the colony', but it was also urgently required for the provision of further water supply. This land was leased by the Chinese for a period of ninety-nine years and is thus due to be returned to China in a little over thirty years time. It is known as the New Territories as distinct from the Colony proper of Hong Kong and Kowloon. At the time of the convention granting this 1898 lease it is estimated that there were about 100,000 inhabitants in its 423 villages. Today this number has grown to about 270,000, the make-up of which still reflects the diverse linguistic grouping of south-east China. The Cantonese are the largest group and the most wealthy. They occupy the better, low-lying agricultural land and are settled in the larger villages, some of which are walled. The Hakka or 'guest people' are the second largest and occupy the higher, less valuable land in the valleys as a rule. Both Cantonese and Hakka stem from early Yueh roots and both are essentially agriculturalists. The third and smaller group is the Hoklo, who probably occupied the land before the Cantonese. They are boat dwellers who today have also settled on the land, particularly on the islands of Cheung Chau and Ping Chau, while the fourth group, the Tanka, are essentially boat people and live their lives with their families afloat.

From 1841 to 1941 decennial reports show a continual rise of population with marked periodic fluctuations occasioned by waves of immigration and exodus to and from China in times of political upheaval or unrest on the mainland. Once firmly established, Hong Kong has from time to time received floods of refugees seeking a temporary haven when trouble brewed across the border. An even more significant cause of these fluctuations has been the influx and exodus of labourers with each boom and recession of international trade. Since 1941, this pattern has changed. In that year, when population was estimated at about $1\frac{1}{2}$ million, Japan occupied the Colony and there followed a mass exodus, which reduced the numbers to about 600,000. The years 1945–6 saw the defeat of Japan and the quick return of the people to the pre-war level. By 1949 the population was assessed at 1,857,000 when it underwent a further large accession as hundreds of thousands of refugees fled into the Colony when the victorious Communist army swept southward. By 1963 numbers had grown to $3\frac{1}{2}$ millions. This amazing growth has several features which distinguish it from the previous pattern. The cessation of hostilities and the establishment of peace have not been attended, as heretofore, by the return of refugees to their homeland. On the contrary, a steady and illegal

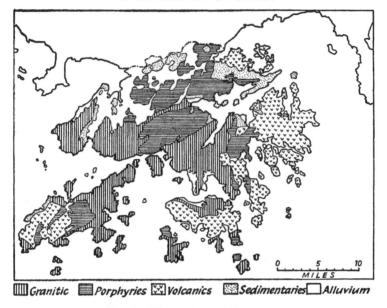

IllGranitic ▦Porphyries ▨Volcanics ▨Sedimentaries▢Alluvium

FIG. 83. The geology of Hong Kong

flow of over 2,000 per month into the colony has continued over the years. In addition to this there has been a phenomenal rise in the natural increase of the population since the end of the war, as the following figures demonstrate:

	Births	Deaths	Natural increase
1949	42,500	13,000	29,000
1953	75,544	18,300	57,244
1958	106,624	21,554	85,070

In 1954 the birth-rate stood at 36·6 $^{0/000}$ and the death rate at 8·5 $^{0/000}$ while only four years later the figures were 38·8 $^{0/000}$ and 7·5 $^{0/000}$ respectively. The reason for this lies very largely in the great improvement in public health and hygiene achieved in spite of the overcrowding. A further reason for population growth has been the great development of industry in the last decade. This aspect will be dealt with later in this chapter. More than 99 per cent of the population of Hong Kong and the New Territories is Chinese.

Hong Kong's geological history reproduces in minature the probable geological sequence in the rest of south-east China. The oldest rocks are either Permian or early Jurassic sediments: the very few fossil remains

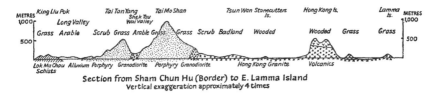

Section from Sham Chun Hu (Border) to E. Lamma Island
Vertical exaggeration approximately 4 times

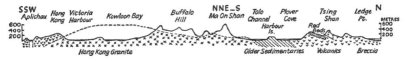

Section from Aplichau to Ledge Point

FIG. 84. Sections across Hong Kong and the New Territories

leave some doubt as to their identity. They were folded along a ENE-WSW axis. These sedimentaries were highly metamorphosed in the Jurasside revolution when there was widespread volcanic outpouring of tuff, lava and agglomerates. A subsequent bathylith gave rise to sills and laccoliths of quartz, porphyry, granite, granite porphyry and swarms of dykes, which are the preponderating and characteristic rocks. A long period of quiescence and erosion was followed first by extensive folding, which affected the whole of south-east China and gave it its NE-SW trend and then by a further outpouring of acid lavas probably during the Miocene. Thus the great bathylith was capped by a thick covering of hard volcanics. Hong Kong granite is singularly susceptible to deep chemical weathering, probably because the felspars break down easily. Where this volcanic cap is pierced by erosion and the granite core exposed, the latter weathers comparatively quickly as the accompanying section reveals.

The relief of Hong Kong and the New Territories is characterized by three broken ranges of steep and often precipitous hills. The three most notable peaks are Tai Mo Shan (3,130 ft.) in the heart of the New Territories, Lan Tao Peak (3,065 ft.) on the large island of Lantao and the precipitous Ma On Shan (2,296 ft.) to the north-east of Kowloon, in which workable iron ore is found. The lines of these hills conform to the NE-SW trend of the geology and occupy about four-fifths of the total area. The only sizeable area of flat land is the alluvial plain in the north-west of the New Territories. The long and very indented coastline is equally rugged, with cliffs of 200 to 600 ft. in the east, where columnar volcanic formations are encountered.

From the full and accurate meteorological records which have been kept for over seventy years in Hong Kong we can get a clear and precise picture

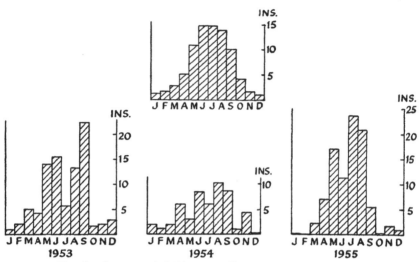

FIG. 85. Average rainfall in Hong Kong and actual rainfall 1953-5

of the climate of this region. Hong Kong lies just within the tropics and on the south-eastern verge of Asia. In consequence it experiences a monsoon rhythm. Easterly winds are prevalent over nine months (September to May) of the year. Only during June, July and August do the main winds come from the south and south-west. It is during these three months that, on average, the heaviest rain falls. A diagram of monthly average rainfall over sixty years gives an almost perfect picture of the monsoon rhythm, but it is false if it leaves the impression of great regularity and invariability. The truth is that considerable variation in rainfall occurs from year to year both in total amount and in monthly distribution as the three diagrams demonstrate.

It is this variability, although not as extreme as in northern China, which constitutes one of the main hazards of the farmer in south China. A failure, such as those of 1954 and 1963, probably means only a single instead of a double crop and may even jeopardize the single crop. The nature of the rainfall varies with the season. Drizzle is characteristic of February and March, damp, cold weather being often experienced at Chinese New Year (February). Summer rain occurs usually in frequent heavy downpours and sometimes in cloud-bursts, which bring phenomenal and unwelcome falls. On 19th July 1926 there was a fall of 3·965 in. in one hour and over 21 in. in 24 hours. In a typhoon the downpour is accompanied by disastrously strong winds.

Hong Kong and the New Territories enjoy a full year's growing period.

The fact that once and only once in seventy years has frost been recorded on the plains is sufficient comment on winter temperatures. Frost and icicles on Victoria Peak, the wealthy residential area of Hong Kong (1,000 to 1,800 ft.) are occasions for headlines in the papers. Although summer temperatures are not spectacularly high—97°F. on 19th August 1900 is the highest recorded—they are uncomfortable on account of the dampness of the atmosphere. Average relative humidity stands between 83 and 85 per cent from April to August. October to December are delightful months with clear blue skies, warm sun and a dry atmosphere.

The natural vegetational cover of the lowlands and most of the hills before human settlement was forest and woodland. This has long since been cleared by felling and burning and now less than 4 per cent is forested. The greater part of this small amount is on Hong Kong island where closer control of cutting can be practised than in the New Territories. The most widely planted trees are the local pine (*Pinus massoniana*), eucalyptus and casuarina. Big efforts are being made at afforestation by the Forestry Department but the difficulties are great, especially in the deeply gullied and eroded granitic 'badlands'. These 'badlands' occupy just over 4 per cent of the Colony and are for the most part entirely devoid of vegetational cover. This is a more important fact than so small a proportion of the total area would suggest, because much of this 4 per cent lies within the catchment basin of the biggest reservoir and therefore seriously affects run-off and water supplies. It is for this reason that the greatest afforestation efforts are concentrated on these lands.[1]

The original woodlands have given place mainly to coarse, tufted grass and low, woody scrub of about 1 ft. in height. These cover 57 per cent of the total area. Another 16 per cent is occupied by taller scrub and bushes. The alluvial plain in the north-west is given over entirely to agriculture and is now entirely devoid of natural cover. Much of the plain has been reclaimed from the sea in the last fifty years, mainly from Deep Bay into which the Shum Chun River empties its silt. Along the shallow coast here are mangrove swamps and marshes.

Climate in Hong Kong is a more important soil-forming agent than mechanical action. The combination of heavy summer rains and high temperatures brings about deep and rapid chemical weathering, particularly in the widespread granites. The resultant soils are clayey and tend to be lateritic. The heavy leaching to which they are subjected every summer results in acid soils everywhere.

[1] T. R. Tregear, *Land Use in Hong Kong and the New Territories* (Hong Kong 1958).

There are about 275,000 persons engaged in the two primary industries of agriculture and fishing. Of these some 150,000 are dependent directly or indirectly on fishing, 60,000 actually living in boats. These are engaged in both deep-sea and inshore fishing, the latter being in grave danger of over fishing. In the last ten years great steps have been made in mechanization of the fleets to meet Japanese competition in this field. In 1958 the catch of marine fish was 44,906 tons and was worth nearly £3 million. In addition to the open sea fisheries, pond pisciculture has developed considerably in the north-west of the New Territories where one acre of pond produces a return twice as great as that of rice paddy.

The following figures give some idea of the distribution of the agricultural land between the various crops.

	acres
Rice Land Two-crop paddy	20,191·97
One-crop brackish water paddy	2,912·15
One-crop upland paddy	248·07
Vegetable land	2,254·87
Orchard	952·36
Field crops	3,479·52

Since rice is the staple food of the Chinese it is not surprising to find that its growth occupies more than two-thirds of the cultivated land. Although most of the paddy is double-cropped, rice production suffices to meet the needs of only a very small proportion of the Colony's population. A great deal is imported from Thailand and China.

Vegetable growing has shown great increase since the Second World War. Before that war only one-fifth of the Colony's needs were met by local growers. Today, in spite of the great growth of population, it is nearly self-sufficing. This is the most intense form of farming, some land producing as much as eight crops a year. The main vegetables grown are: white cabbage, flowering cabbage, leaf mustard cabbage, turnips, Chinese kale and Chinese lettuce. Field crops include sweet potatoes, taro, yams, sugar cane and groundnuts.

Both the fishermen and vegetable growers have received great encouragement since the war by the establishment of the Fish Marketing Organization and the Vegetable Marketing Organization, which, although receiving help from government personnel, are non-government bodies. They are designed to organize the collection, sorting and fair marketing of the produce, eliminating private middlemen's charges, which formerly were exorbitant. In addition to marketing, the organizations have social and

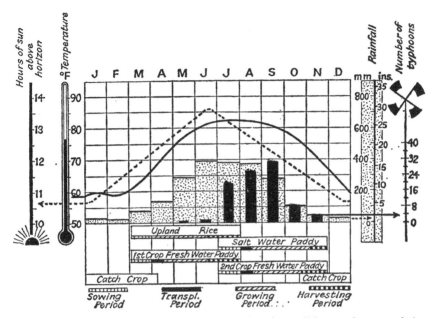

FIG. 86. Relation between Hong Kong climatic conditions and crop period of rice

educational activities. Co-operative societies, assisting in buying, selling and credit are growing and flourishing in both the fishing and farming communities. Surprisingly, fruit farming in the past has received little attention. It is now increasing. Papaya, guava, Chinese lime, tangerine, lychee, *wong pei* and peach are the main fruits grown. Pineapple plantations are increasing and regaining some of their former popularity. Tea, which was grown widely on the hill slopes in the 1870's and 1880's receives no attention whatever today. Agriculture in all its branches, including livestock, receives much advice, help and demonstration from an active Government Agricultural, Fisheries and Forestry Department, and encouragement and financial assistance through such philanthropic bodies as the Joseph Trust Fund and the Kadoorie Agricultural Aid Association.

However, Hong Kong's interest and *raison d'être* has been from the outset not in agriculture but in commerce. For one hundred years, between 1841 and 1941 exactly, as a free port, having an almost complete absence of import duties, it was one of the great entrepôts of the world. Its large natural harbour was always full of ships of all nations, collecting and distributing the raw materials of China and the manufactured articles of the Western world. Except for flourishing shipbuilding and ship repair works,

317

industry was of little importance. With the Japanese occupation of the port in 1941, all international commerce was brought to an abrupt halt; until 1946 it served only Japanese war needs. Trade recovery was quick when peace returned and was broadly of the pre-war pattern. By 1951 a new peak was reached when imports were valued at £300 million and exports at £275 million. Then followed in the next four years an equally rapid and disastrous slump, when, with the outbreak and continuance of the Korean War the United Nations embargo was placed on a long list of strategic goods entering China. By 1954 imports had fallen by about £80 million and exports by more than £100 million. Hong Kong's answer to this unhappy state has been revolutionary, redounding to its energy and resilience. In a matter of less than a decade it has largely changed the basis of its economic life from reliance on commerce to industry. The following table conveys some idea of the magnitude of the change which has taken place.

Growth of Industry in Hong Kong 1948–58

Workers	1948	1958
Metal workers	9,914	24,342
Shipbuilding and repairs	9,729	10,049
Food manufacture	3,308	6,921
Plastic ware	33	8,024
Rubber products	4,427	8,788
Cotton weaving	6,488	15,870
Cotton spinning	1,755	12,613
Wearing apparel	1,196	25,602

Most of the industrial development has taken place in and around Kowloon. Many new centres have sprung up overnight. For example, Tsun Wan in 1952 was a village of about 5,000, where a small camphor industry had once been carried on. Today it is a hive of activity with a population of more than 80,000. It will be seen from the above figures that, while shipbuilding remains an important industry, it has not grown significantly. It is in the sphere of light industry that the startling changes have taken place, particularly in textiles. The number of workers engaged in spinning, weaving and garment-making rose from 9,439 in 1948 to 54,085 in 1958. Modern machinery has been installed; new skills have been quickly acquired; the quality of output is good. It is not surprising that Lancashire has felt the competition from this quarter. Many other new industries have been started. Enamel and aluminium-ware, plastic goods, electric torches and batteries, vacuum flasks, paints, boots and shoes,

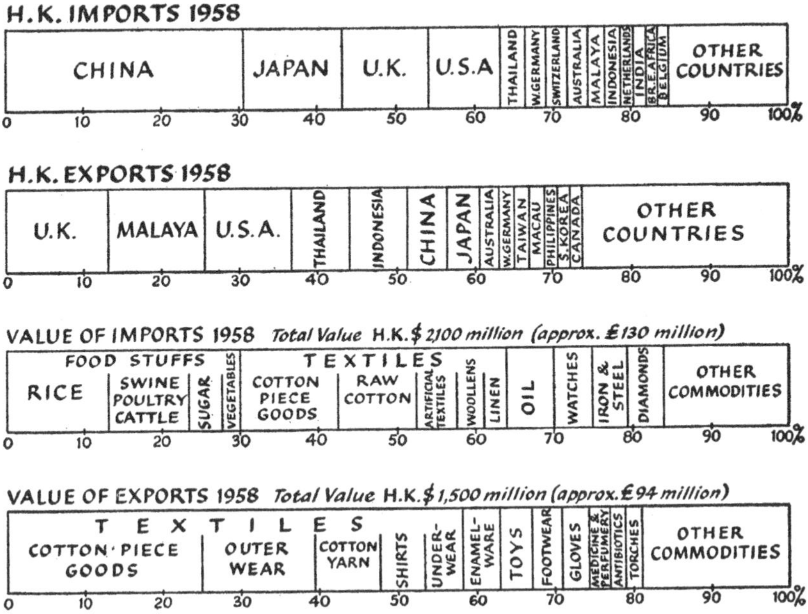

H.K. IMPORTS 1958

CHINA	JAPAN	U.K.	U.S.A	THAILAND	W.GERMANY	SWITZERLAND	AUSTRALIA	MALAYA	INDONESIA	INDIA	NETHERLANDS	BRE.AFRICA	BELGIUM	OTHER COUNTRIES

```
0    10    20    30    40    50    60    70    80    90   100%
```

H.K. EXPORTS 1958

U.K.	MALAYA	U.S.A.	THAILAND	INDONESIA	CHINA	JAPAN	AUSTRALIA	W.GERMANY	TAIWAN	MACAU	PHILIPPINES	S.KOREA	CANADA	OTHER COUNTRIES

```
0    10    20    30    40    50    60    70    80    90   100%
```

VALUE OF IMPORTS 1958 *Total Value H.K.$ 2,100 million (approx. £130 million)*

FOOD STUFFS				TEXTILES					OIL	WATCHES	IRON & STEEL	DIAMONDS	OTHER COMMODITIES
RICE	SWINE POULTRY CATTLE	SUGAR	VEGETABLES	COTTON PIECE GOODS	RAW COTTON	ARTIFICIAL TEXTILES	WOOLLENS	LINEN					

```
0    10    20    30    40    50    60    70    80    90   100%
```

VALUE OF EXPORTS 1958 *Total Value H.K.$ 1,500 million (approx.£94 million)*

TEXTILES				SHIRTS	UNDER-WEAR	ENAMEL-WARE	TOYS	FOOTWEAR	GLOVES	MEDICINE & PERFUMERY	ANTIBIOTICS	TORCHES	OTHER COMMODITIES
COTTON·PIECE GOODS	OUTER WEAR	COTTON YARN											

```
0    10    20    30    40    50    60    70    80    90   100%
```

FIG. 87. Diagram of the foreign trade of Hong Kong

leather goods and many others figure in the list. Most of the factories are small by western standards; only two employ more than 2,000 workers. In addition to this more organized industry it is estimated that about 150,000 persons are engaged in traditional Chinese handicrafts, such as embroidery and drawn-thread work, brocade piece-goods, wood and ivory carving.[1]

Hong Kong's mineral wealth is not great. There are small deposits of lead, silver, wolfram and tin associated with the many dykes and intrusions. These are worked in a small way as also is some graphite on the Brothers Islands to the north of Lan Tao. Kaolin is worked in the west and used both in the local ceramic industry and exported. Iron ore workings on Ma On Shan are in the hands of a Japanese concession, whose exports to Japan are valued at about H.K.$4 million or £250,000 annually.

From earliest days the Colony has been in trouble over water supplies. Population has constantly outrun engineering efforts as reservoir after reservoir has been built. In an endeavour to meet the needs of growing industry and the rapid rise in population during the last ten years, two very large dams with large catchment areas at Tai Lam Chung and Shek Pik on Lan Tao have been built. Although this trebled the water supply, it

---

[1] E. Szczepanik, *The Economic Growth of Hong Kong* (London 1958), gives a full treatment of the subject.

has been found insufficient and a resort has had to be made to supplies from the mainland of China.

Even in so short a description of Hong Kong as this, some comment on the refugee problem is called for, since it is unique. Of Hong Kong's 3 million inhabitants, between 700,000 and 800,000 have entered from the mainland in the last twelve years either for political reasons, fleeing from a government which is feared, or for economic and social reasons, or both. Some of the refugees have been wealthy but most are indigent; some are skilled but most are unskilled. They have not been herded into camps as in Europe but have mingled with, and are largely indistinguishable from, the rest of the population as the Hambro U.N. Commission discovered. Many have had friends and relations in the Colony but very many more have not, and have squatted in miserable shacks on the hillsides around the two cities of Victoria and Kowloon. Such an influx has imposed enormous burdens on the authorities. Disastrous fires sweep through these squatter settlements. One alone on Christmas Eve 1954 rendered 60,000 homeless in one night. Overcrowding in some parts of the cities is very great. In Wan Chai, for example, there are as many as 2,000 people to the acre. To meet this problem the government has built large numbers of huge tenement blocks. Other problems of unemployment, underemployment, malnutrition, sanitation and not least education all follow in this train. In spite of all, the Colony manages to present a gay, busy, bustling yet orderly and cleanly community, full of that *je nao* (hot noise) which most Chinese love.

12. MACAU

Macau is a Portuguese colony situated on the right bank at the mouth of the Pearl River estuary, some 35 miles WSW of Hong Kong. It consists of a small peninsula and two islands, Taipa and Coloane, having a total area of only 10–11 sq. miles.

In 1516 the Portuguese landed on the small peninsula and there established their first trading station. It remained in their possession but under nominal Chinese control until in 1849 the Portuguese declared it to be their own territory. Although this declaration was confirmed by treaty in 1887, it has continued a matter of disputation ever since. It became the chief centre of European trade with China and so remained until the cession of Hong Kong to the British. Thereafter its importance as an entrepôt steadily and rapidly declined. Today it is a sleepy, picturesque but decadent township in striking contrast to the bustle, noise and activity of Hong Kong.

Apart from fishing, which provides a livelihood for about a quarter of its 180,000 inhabitants, and the manufacture of firecrackers, the town is given over to the tourist trade. It is a favourite week-end resort of the people of Hong Kong with which it is connected by an efficient and comfortable ferry service. There are many hotels and a rather down-at-heel gambling casino. Formerly it was a notorious opium-smuggling centre. Today it is one of the main routes through which refugees from Communist China pass in their endeavour to reach Hong Kong.

There are less than 3,000 Portuguese now resident in Macau.

CHAPTER FIVE

The Unity of China

The detailed examination of comparatively small regions, which has been the subject of the preceding chapters, may have tended to obscure or at least to have pushed into the background the broad differences that exist between north and south China, the dividing line of which is marked by the Tsinling. We must return, therefore, to these differences which have been noted by all geographers who have written on China, notably George B. Cressey and L. Dudley Stamp, both of whom have emphasized the contrast by detailed tabulation.

We have seen the physical dissimilarities such as the tropical and subtropical temperatures of the south and the continental temperature ranges of the north; the greater reliability of rainfall in the south, the long growing season in the south and the short one in the north; the contrast of the acid soil of the south and alkaline of the north; the grasslands and sparse woodlands of the north and the sub-tropical natural forest of the south. All these are natural differences which will remain unaltered by man in the foreseeable future and which will continue to exert a powerful influence on his activities. Some modification of soils may be possible and the distribution of vegetation may be changed somewhat by afforestation programmes, but by and large climate will remain the main arbiter of man's activities in these fields.

It is the clear and stated policy of the present communist government to bring the country, formerly so fragmented, into a close political unity. To do this it is necessary that it should iron out as many of the differences between north and south as possible, and it is using all means in its power—political, economic, social and educational—to achieve this end.

The larger farms and the larger fields of the north stand in marked contrast to the necessarily smaller paddy fields of the south, which, following the contours of the hillsides, are often not more than a yard or two wide. Here two or three modern trends are discernible. Wheat is on the whole a better cereal than rice, having better food values. While the wheat crop per acre is not as heavy as that of rice, it is rather less dependent on climatic conditions, it is far less demanding of labour in its cultivation and, by its larger fields, it lends itself to large scale farming, such as is envisaged in the commune, far more easily than rice. For these reasons wheat cultivation commends itself to China's planners. Nevertheless a great deal of education, both of the farmer and the consumer, will have to be done if the wheat line is to be extended southward appreciably. The Chinese, whether northerner or southerner, has a marked predilection for rice as his staple diet if he can get it. This is particularly true the farther south one goes. In any case it would be tempting a meteorological providence to attempt to push wheat cultivation very far south.

Water conservation work on the Hwang Ho and the Hwai Ho will, if successful, reduce the menace of flood which has hung so heavily over the north for so many centuries and so will reduce a long-standing contrast between north and south.

Perhaps more than any other single factor the development of communications during the last decade had done most towards achieving a unity and producing a merging of north and south, for only when people can communicate freely and easily in trade and ideas can there be a real integration. Throughout the centuries the northerner and southerner have maintained distinctive characteristics which are far clearer than those, for example, distinguishing the English and Welsh. There are differences of stature and of temperament, the northerner *Han Jen* being markedly taller, slower and more phlegmatic than the shorter, more volatile and revolutionary *T'ang Jen* of the south. The northerners also are of more uniform race, evincing more Mongolian characteristics than the southerners, who have a large admixture of the many tribes of the south. Ever since the south was occupied 2,000 years ago there has always been rivalry, not to say hostility, between the regions. This mutual distrust still exists and any government must have regard to it. How aware the present government is of this fact was evinced by the speed with which they sent so many of their best administrators to the south when they unexpectedly gained such rapid control of the country in 1949. These differences will tend to disappear slowly as there is more and more free intercourse and easy movement.

The development of communications is also already playing an important part in bringing about a uniformity of speech and ideas, which is imperative if north and south are to be brought together. While the same written language runs throughout the whole land, the north is essentially Mandarin-speaking and the south mingles Cantonese with a vast number of dialects. Mandarin and Cantonese are so different as to be two distinct languages. The People's Government is endeavouring to develop a common speech as well as a common written script by making the study of a modified form of Mandarin, known as *jen min hua*, compulsory in all schools. How long and how effective their present measures will be remains to be seen. Their difficulties in training an adequate teaching staff for all the education they wish to undertake are greater probably than those of any other country in the world. Great energy is also being expended in the endeavour rapidly to make the whole country literate. To this end much research is being put into simplifying Chinese characters, i.e. ideographs, and into trying to find a romanization which will present the meaning with the same clarity as the character—a task which up to the present has defied all attempts.

Indoctrination of the communist Marx-Lenin-Mao creed continually plays its part in the endeavour to knit north and south into one thinking whole within the space of one or two decades.

Whatever the differences of north and south, there are certain characteristics, which are common to both. The Chinese are a practical, hardworking, persistent and philosophical people, capable of meeting and sustaining the bitter and exacting hardships which the present revolution towards an industrial state will inevitably continue to demand of them for many years to come.

Bibliography

Andersson, J. G., *Children of the Yellow Earth*, Routledge & Kegan Paul, 1934; 'Researches into the Pre-history of the Chinese', *Bulletin of the Museum of Far Eastern Antiquities*, 1943.

Barbour, G. B., 'Recent Observations on the Loess of North China', *Geographical Journal*, 1935, Nos. 54–64; 'Physiographic History of the Yangtze', *Geographical Journal*, 1936; 'The Loess Problem of China', *Geographical Magazine*, 1930.

Belden, W. and Salter, M., 'Iron Ore Resources of China', *Economic Geography*, 1935.

Biot, E., 'Mémoire sur les Colonies Militaires et Agricoles des Chinois', *Journal Asiatique*, 1850.

Bishop, C. W., 'The Rise of Civilisation in China with reference to its Geographical Aspects', *Geographical Review*, 1932; 'Beginnings of North and South in China', *Pacific Affairs*, Vol. 7, No. 297, 1934; 'The Geographical Factor in the Rise of Chinese Civilisation', *Geographical Review*, 1922.

Borchert, J. R., 'A New Map of the Climates of China', *Annals of the Association of American Geographers*, No. 169, 1947.

Buck, J. Lossing, *Land Utilization in China*, Commercial Press, Shanghai, 1937; *Chinese Farm Economy*, Commercial Press, Shanghai, 1937.

Buxton, L. H. D., *China, The Land and the People*, Oxford University Press, 1929.

Cable, M. and French, F., *The Gobi Desert*, Hodder & Stoughton, 1946.

Carles, W. R., 'The Yangtze Kiang', *Geographical Journal*, Vol. 12, No. 225, 1898.

Chandrasekhar, S., *China's Population*, Hong Kong University Press, 1959.

Chang Chih-yi, 'Land Utilization and Settlement Possibilities in Sinkiang', *Geographical Review*, No. 39, 1949.

Chang Sun, 'The Sea Routes of the Yuan Dynasty, 1260–1341', *Acta Geographica Sinica*, Vol. 23, 1959.

Chavannes, E., *Les Mémoires Historiques de Ssu-Ma Chien*, Vol. 3, Leroux, Paris, 1898.

Chen, Kenneth, 'Early Expansion of Chinese Geographical Knowledge', *T'ien Hsia Monthly*, Vol. II, 52, 1940–1.

Cheng Te Khun, 'An Introduction to Chinese Civilization', *Orient*, August-October 1950; 'Short History of Szechwan', *Journal of the West China Research Society*, No. 16, 1945.

Ch'i Ch'ao-ting, *Key Economic Areas in Chinese History, as revealed in the Development of Public Works for Water Control*, Allen & Unwin, 1936.

Chiang Knoh, 'The Tung Region of China', *Economic Geography*, No. 418, 1943.

Chu Co-ching, 'The Circulation of Atmosphere over China', *Memorandum of Meteorology*, No. 4, Nanking; 'The South-east Monsoon and Rainfall in China', *Journal of the Chinese Geographical Society*, Vol. 1, No. 1, 1934.

Codrington, K. de B., 'A Geographical Introduction to the History of Central Asia', *Geographical Journal*, Vol. 104, 1944.

Collins, W. F., *Mineral Enterprise*, Probsthain, 1918.

Cornish, Vaughan, *The Great Capitals*, Methuen, 1922.

Cottrell, L., *The Tiger of Ch'in*, Evans 1962, Pan Books, 1964.

Creel, G. H., *Studies in Early Chinese Culture*, Waverly, Baltimore, 1937, Routledge, 1938; *The Birth of China*, Waverly, Baltimore, 1937.

Cressey, G. B., *China's Geographic Foundations*, McGraw-Hill, 1934; *Land of the 500 Million*, McGraw-Hill, 1955; 'Land Forms of Che-Kiang', *Annals of the Association of American Geographers*, No. 259, 1938; 'Foundations of Chinese Life', *Economic Geography*, No. 95, 1939; 'The Ordos Desert of Inner Mongolia', *Denison University Bulletin*, Vol. 33, No. 8.

Davies, R. M., *Yunnan, the Link between India and the Yangtze*, Cambridge, 1909.

Davis, S. G., *Hong Kong in its Geographical Setting*, Collins, 1949; *The Geology of Hong Kong*, Government Printer, Hong Kong, 1952.

Deasy, 'Tung Oil Production and Trade', *Economic Geography*, No. 260, 1940.

Donnithorne, A., 'Economic Development in China', *The World Today*, Vol. 17, No. 4, 1961.

Drake, F. S., 'The Struggle for the Tarim Basin in the Later Han Dynasty', *Journal of the Royal Asiatic Society*, North China Branch, 1935; 'Mohammadanism in the Tang Dynasty', *Monumenta Serica*, No. 8, 1943.

East, W. G. and Spate, O. H. K., *The Changing Map of Asia*, Methuen, 1953.

Eberhard, *A History of China*, Routledge & Kegan Paul, 1952.

Fei Hsiao-tung, *Peasant Life in China*, Routledge & Kegan Paul, 1947.

Fei Hsiao-tung and Chang Chih-i, *Earthbound China, A Study of Rural Economy in Yunnan*, University of Chicago Press, 1945.

Fenzel, G., 'On the Natural Conditions affecting the Introduction of Forestry in the Province of Kwangtung', *Lingnan Science Journal*, No. 7, 1929.

Ferguson, J. C., 'The Southern Migration of the Sung Dynasty', *Journal of the Royal Asiatic Society*, North China Branch, No. 55, 1924.

Fitzgerald, C. P., 'The Yunnan-Burma Road', *Geographical Journal*, Vol. 95, No. 161, 1940; 'The Tiger's Leap', *Geographical Journal*, Vol. 98, No. 147, 1941; 'The Tali District of W. Yunnan', *Geographical Journal*, Vol. 99, No. 50, 1942; 'The Northern Marches of Yunnan', *Geographical Journal*, Vol. 102, No. 49, 1943; *Revolution in China*, Cresset Press, 1952; *Flood Tide in China*, 1958.

Foord, E., 'China and the Destruction of the Roman Empire', *Contemporary Review*, Vol. 94, No. 207, 1908.

Fox, R., *Genghis Khan*, Lane, 1936.

Gamble, S. D., *Ting Hsien: a North China Rural Community*, Institute of Pacific Relations, 1954.

Geil, W. E., *The Eighteen Capitals of China*, Constable, 1911; *The Great Wall of China*, Constable, 1909.

Gherzi, E., *Climatological Atlas of East Asia*, Shanghai, 1944.

Giles, H. A., *The Travels of Fa Hsien*, Cambridge University Press, 1923.

Ginsburg, N. S., 'Ch'ing-tao: Development and Land use', *Economic Geography*, No. 181, 1948; 'China's Changing Political Geography', *Geographical Review*, No. 102, 1952.

Gourou, P., *The Tropical World*, Longmans, 1954.

Grabau, A. W., *The Stratigraphy of China*, Peking, 1925.

Granet, M., *Chinese Civilization*, Routledge, 1930.

Grazdanzev, A. J., 'Manchuria, 1945: an Industrial Survey', *Pacific Affairs*, Vol. 18, December 1945.

Green, F., *The Wall has Two Sides*, Cape, 1961.

Gregory, J. W., 'Is the Earth drying up?' *Geographical Journal*, 1914.

Hanson-Lowe, J., 'Structure of the Lower Yangtze Terraces', *Geographical Journal*, 1939.

Hare, F. K., *The Restless Atmosphere*, Hutchinson, 1953.

Herrmann, A., *Atlas of China*, Harvard-Yenching Institute, Cambridge, Massachusetts, 1935.

Hirth, F., 'The Story of Chang Ch'ien, China's Pioneer in West Asia', *Journal of the American Oriental Society*, 1917.

Hitch, M., 'The Port of Tientsin and its Problems', *Geographical Review*, 1935.

Ho Ping-ti, *Studies on the Population of China*, 1368–1953, Harvard University Press, Cambridge, Massachusetts, 1959; *Hong Kong Annual Report*, Government Press, Hong Kong.

Hsia, R., *Economic Planning and Development in Communist China*, Institute of Pacific Relations, New York, 1955.

Hu, C. Y., 'Land Use in the Szechwan Basin', *Geographical Review*, 1947.

Hu Huan-yong, 'A Geographical Sketch of Kiangsu Province', *Geographical Review*, 1947; 'A New Cotton Belt in China', *Economic Geography*, 1947.

Hudson, G. F., *Europe and China*, Arnold, 1931.

Hughes, E. R., *The Invasion of China by the Western World*, Black, 1937.

Hughes, R. H., 'Hong Kong: an Urban Study', *Geographical Journal*, 1951.

Hughes, T. J. and Luard, D. E. T., *The Economic Development of Communist China*, 1949–1958, Royal Institute of International Affairs, Oxford, 1959.

Hung Fu, 'The Geographic Regions of China and their Sub-divisions', *International Geographical Union Proceedings*, 1952.

Huntington, E., *The Pulse of Asia*, Boston, 1910.

James, H. F., 'Industrial China', *Economic Geography*, 1929.

Jen, M. N., 'Agricultural Landscape of S.W. China', *Economic Geography*, 1948.

Jones, F. O., 'Tukiangyien: China's Ancient Irrigation System', *Geographical Review*, 1954.

Kendrew, W. G., *Buxton's China*. Chapter 14 on Climate, Oxford University Press, 1929.

King, F. H., 'The Wonderful Canals of China', *National Geographic Magazine*, 1912.

King, F. H., *Farmers of Forty Centuries*, Harcourt Brace, 1926.

Kingdom-Ward, F., 'Tibet as a Grazing Land', *Geographical Journal*, 1947.

Kirby, E. Stuart, *Introduction to the Economic History of China*, Allen and Unwin, 1954.

Komroff, M. (Ed.), *The Journey of Friar John of Pian de Carpini to the Court of Kuyak Khan, 1245–1247*, Jonathan Cape, 1929.

Kuo Ping-chia, *China: New Age and New Outlook*, Penguin Books, 1960.

Kuo Ts'ung-fei, 'Brief History of the Trade Routes between Burma, Indo-China and Yunnan, *T'ien Hsia Monthly*, Vol. 12, 1940–41.

Lapwood, R. and N., *Through the Chinese Revolution*, Spalding & Levy, 1954.

Latourette, K. S., *The Chinese, Their History and Culture*, Macmillan, 1956.

Lattimore, O., *Inner Asian Frontiers*, Oxford University Press, 1940; 'Inner Asian Approach to the Historical Geography of China', *Geographical Journal*, 1947; 'Origins of the Great Wall of China', *Geographical Review*, 1937; 'Caravan Routes of Inner Asia', *Geographical Journal*, 1928; 'Chinese Colonization in Manchuria', *Geographical Review*, 1932.

Lee, H. K., 'Korean Migrants in Manchuria', *Geographical Review*, 1932.

Lee, J. S., *The Geology of China*, Murby, 1939.

Legge, J., *The Chinese Classics* (5 Vols), London, 1861–72.

Leong, Y. K. and Yao, L. K., *Village and Town Life in China*, Allen, 1915.

Li Choh-Ming, *The Statistical System of Communist China*, University of California, 1962.

Lin En-lau, 'The Ho-si Corridor (Kansu pan-handle)', *Economic Geography*, 1952.

Lin Yu-tang, *My Country and My People*, Heinemann, 1948.

Little, A., *The Far East*, Oxford University Press, 1905.

Lu, A., *The Climatological Atlas of China*, Central Weather Bureau, Nanking, 1946.

Mallory, W. H., *China, Land of Famine*, American Geographical Society, 1926.

Mason, I., 'The Mohammedans of China: when and how they first came', *Journal of the Royal Asiatic Society*, North China Branch, 1929.

Ma Yung-chi, *General Principles of Geographical Distribution of Chinese Soil*, Peking, 1956.

Miller, A. A., *Climatology*, Methuen, 1946.

Mordes, F., *The Revolt in Tibet*, Macmillan, 1960.

Moyer, R. T., 'Agricultural Soils in a Loess Region of North China', *Geographical Review*, 1936.

Min Tieh, 'Soil Erosion in China', *Geographical Review*, 1941.

Muzaffer Er Selçuk, 'Iron and Steel Industry in China', *Economic Geography*, 1956.

Needham, J., *Science and Civilization in China*, Vols. 1–3, Cambridge University Press, 1954.

Norin, E., 'Quaternary Climatic Changes within the Tarim Basin', *Geographical Review*, 1952.

Orchard, 'Shanghai', *Geographical Review*, 1936.

Pelliot, P., 'Note sur les Anciens Itinéraires Chinois dans l'Orient Roman', *Journal Asiatique*, 1921.

Purcell, V., *The Chinese in South-east Asia*, Oxford University Press, 1951.

Radhakkrishnan, S., *India and China*, Hind Kitab, Bombay, 1947.

329

Ramage, C. S., 'Evapotranspiration Measurements made in Hong Kong.' First Report, Technical Notes No. 7, Royal Observatory, Hong Kong, 1953.

Read, T. T., 'Economic-Geographic Aspects of China's Iron Industry', *Geographical Review*, 1943.

Regis, G., 'Developments in Chinese Agriculture', *Far East Trade*, Vol. 17.

Richardson, H. L., 'Szechwan during the War', *Geographical Journal*, 1945.

von Richthofen, 'On the Mode of Origin of the Loess', *Geographical Magazine*, 1882.

de Riencourt, A., *The Soul of China*, Cape, 1959.

Rogers, A., 'Manchurian Iron and Steel Industry and its Resource Base', *Geographical Review*, 1948.

Roxby, P. M., 'The Distribution of Population in China', *Geographical Review*, 1935; 'The Expansion of China', *Scottish Geographical Magazine*, 1930; 'China as an Entity: the Comparison with Europe', *Geography*, 1934; 'The Terrain of Early Chinese Civilisation', *Geography*, 1938.

Schoff, W. H., 'Navigation to the Far East under the Roman Empire', *Journal of the American Oriental Society*, 1917; 'Some Aspects of the Overland Oriental Trade at the Beginning of the Christian Era', *Journal of the American Orient Society*, 1915.

Schomberg, R. C. F., 'The Habitability of Chinese Turkestan', *Geographical Journal*, 1932.

Shabad, T., *China's Changing Map*, Methuen, 1956.

Shen, T. H., *Agricultural Resources of China*, Cornell University Press, 1951.

Shu Tan Lee, 'Delimitation of the Geographic Regions of China', *Annals of the Association of American Geographers*, 1947.

Smith, W., *Iron and Coal in China*, Liverpool University Press, 1926.

Snow, E., *Red Star over China*, Gollancz, 1962; *The Other Side of the River*, Gollancz, 1962.

Spencer, J. E., 'Salt in China', *Geographical Review*, 1935.

Stamp, L. D., *Asia*, Methuen, 1957.

Stein, Sir A., *Innermost Asia:* 2 vols, Oxford University Press, 1928; 'Innermost Asia: its Geography as a Factor in History', *Geographical Journal*, 1925.

Stevenson, P. H., 'Notes on the Human Geography of the Chinese-Tibetan borderland', *Geographical Review*, 1952.

Stewart, J. R., 'Manchuria, the Land and its Economy', *Economic Geography*, 1932.

Szczepanik, E., *The Economic Growth of Hong Kong*, Oxford University Press, 1958.

Tawney, R. H., *Land and Labour in China*, Allen & Unwin, 1932.

Teggart, F. J., *Rome and China: a Study of Correlations in Historical Events*, University of California Press, 1939.

Teng Tse-hui, *Report on the Multiple Purpose Plan for Permanently Controlling the Yellow River*, Peking, 1955.

Thompson, B. W., 'An Essay on the General Circulation of the Atmosphere over S.E. Asia', *Quarterly Journal of the Royal Meteorological Society*, Vol. 77, 1951.

Thorp, J., *Geography of the Soils of China*, 1935.

Tregear, T. R., *Land Use in Hong Kong and the New Territories*, Geography Publications, Hong Kong University Press, 1958; 'Shih Hui Yao—A Chinese River Port with a Future', *Geography*, April 1954.

Tregear, T. R. and Berry, L., *The Development of Hong Kong and Kowloon as told in Maps*, Hong Kong University Press and Macmillan, 1959.

Trewartha, G. T., 'Chinese Cities: Numbers and Distribution'; 'Chinese Cities: Origins and Functions', *Annals of the Association of American Geographers*, 1951 and 1952; 'Ratio Maps of China's Farms and Crops', *Geographical Review*, 1938

Trewartha, G. T. and Shou Jen Yang, 'Notes on Rice Growing in China', *Annals of the Association of American Geographers*, 1948.

Tsui Yu-wen, 'Problems of Soil-Preserving Vegetation in the middle reaches of the Hwang-ho', *Academia Sinica*, 1957.

Tu Chang-wang, 'China's Weather and World Oscillation', *National Research Institute of Meteorology*, Vol. 11, 1937.

Tu Chang-wang and Hwang Sze-sung, 'The Advance and Retreat of the Summer Monsoon in China', *Meteorology Magazine*, Vol. 18, 1944.

Valkenburg, S., 'Agricultural Regions of Asia: China', *Economic Geography*, 1934.

Waln, N., *The House of Exile*, Cresset Press, 1933.

Wang Kung-ping, 'Mineral Resources of China, with special reference to Non-Ferrous Metals', *Geographical Review*, 1944.

Ward, F. K. 'Tibet as a Grazing Land', *Geographical Journal*, 1947.

Waring, H. W. A., *Steel Review*, British Iron and Steel Federation, 1961.

Watson, W., *China*, Thames & Hudson, 1961.

Whyte, Sir F., *China and Foreign Powers*, Royal Institute of International Affairs, Oxford, 1927.

Wiens, H. J., *China's March towards the Tropics*, Shoe String Press, Hampden, U.S.A., 1954. 'The Shu Tao or Road to Szechwan', *Geographical Review*, 1949.

Wilson, E. H., *A Naturalist in West China*, Methuen, 1913.

Wittfogel, K. A., *Chinese Society and the Dynasties of Conquest* (in *China* Ed. McNair), University of California Press, 1946.

Yang, M. C., *A Chinese Village; Taitou, Shantung Province*, Routledge & Kegan Paul, 1946.

Yetts, W. P., 'Links between China and the West', *Geographical Review*, 1926.

Yule, Sir H., *The Book of Ser Marco Polo, the Venetian* (Ed. H. Cordier), Murray, London, 1920.

PERIODICALS

The China Quarterly
Journal of Asian Studies (formerly *Far Eastern Quarterly*)
Pacific Affairs
Pacific Viewpoint
Far East Trade
Far Eastern Economic Review
Translations from China Mainland Press, Selections from China Mainland Magazines and Current Background, American Consulate General, Hong Kong
Communist Publications, Peking
    Peking Review (weekly)
    China Reconstructs (monthly)

Glossary

CHINESE	EXPLANATION AND EXAMPLES
tung	east
hsi or *si*	west The Chinese refer to the points of the compass in
nan	south the order E.W.S.N. and not N.S.E.W. as in English
pei or *peh*	north
shang	above, upon, on *Shanghai* – on the sea
hsia	below, lower
ta	big *Ta Kiang* – Yangtze; *ta feng* – a great wind, typhoon
hsiao	little
ch'ang	long *Ch'ang Kiang* – Yangtze, i.e. long river
tuan	short
kiang	river *Yangtze Kiang*; *Si Kiang*
ho	river *Hwang Ho*
shui	water, also used for river
feng	wind *Ta feng*
fengshui	The Spirit of Wind and Water, supposed to reside in the hills of a locality. Much used in geomancy for the auspicious siting of graves and buildings
shan	mountain *Shantung* and *Shansi*; *Nanshan*
ling	hill or mountain *Tsinling*; *Nanling*
k'ou	mouth *Hankow* – mouth of the Han River
men	gate, pass *Hengyangmen*
t'ai	very, excessive *T'ai P'ing Yang* – Pacific Ocean
p'ing	peaceful, level; title of the *T'ai P'ing* rebels
chou (*chow*)	island *Hangchow*
sha	sand *Shatien*
t'ien	field
kang (*kong*)	port
hsiang (*hong*)	fragrant *Hong Kong*
hei	black
lung	dragon *Heilung Kiang* – Black Dragon River
hung	red
hwang	yellow *Hwang Ho*

333

king	capital *Nanking*; *Peking*
k'ai	open
feng	to seal up, blockade *Kaifeng*
an	peace *Sian* – western peace
chou	sub-division of a province – a county
hsien	a district, formerly a sub-division of a prefecture
t'ien	heaven *Tientsin*

Conversion Tables

li	one-third mile
mow	0·06 hectare; 0·165 acre
chin or *catty*	one and one-third lb.
picul	100 *chin*; 133 lb.
tan	100 *chin* or one man's load
tael (Haikwan)	583·3 grains silver
yuan	6·857 *yuan* equal £1 sterling
1 *kilometre*	0·621 miles
1 *sq. metre*	10·763 sq. ft.; 1·196 sq. yd.
1 *sq. yd.*	0·836 sq. metres
1 *sq. ft.*	0·093 sq. metres
1 *hectare*	2·471 acres
1 *acre*	0·405 hectares
1 *sq. mile*	2·589 sq. kilometres
1 *sq. km.*	0·386 sq. miles
1 *lb.*	0·454 kilograms
1 *kilogram*	2·205 lb.

Index

Page numbers in italics refer to text figures